The Collected Angers

AN OBVIOUS GRIFT TO SHAKE DOWN
MY MOST LOYAL READERS

(SHITTY PULP EDITION)

MIKE MONTEIRO

MORE FROM MULE BOOKS

DESIGN IS A JOB, THE NECESSARY 2ND EDITION
Mike Monteiro

JUST ENOUGH RESEARCH, 2024 EDITION
Erika Hall

RUINED BY DESIGN
Mike Monteiro

upcoming:

THE BUSINESS MODEL IS THE GRID
Erika Hall

CONVERSATIONAL DESIGN, 2ND EDITION
Erika Hall

PULP TRASH SPECIAL EDITION

Copyright © 2025 by Mike Monteiro
All rights reserved. All wrongs reversed.
ISBN: 979-8-9893587-9-3

Editors: Mandy Keifetz, Killian Pirarro, Ani King, Erika Hall
Composition: Miguel Mosquito
Cover Design: Miguel Mosquito

Mule Books
601 Minnesota St. #122 SF, CA 94107
mulebooks.com

Publishing is dead. Good riddance. Do it yourself.

TABLE OF CONTENTS

THE RIGHTEOUS ANGERS
My People Were In Shipping 11
On My Second Birthday We Landed On The Moon 19
Design's Lost Generation 25
Hope Is Not Enough 36
Ethics Offsets Are Bullshit 41
Ethics And Paying Rent 46
It Might Be Time To Start Flipping Tables 50
Ignorance Beats Empathy 56
Politics Is The Design Problem Of Our Times 59
How To Fight Fascism 61
In Praise Of The Ak-47 66
The Happiest Article You Will Ever Read About Design Ethics... 69
Doing The Right Thing The Wrong Way 74
Beware The Judas Goat 77
The Best Thing We Can Do For This Planet Is Die 81
The Road Back 84

THE TECH BRO ANGERS
One Person's History Of Twitter, From Beginning To End 95
How To Explain To Your Children That You Work At Facebook .103
The People Vs Donald Trump Vs Twitter 108
Facebook Isn't A Community, It's A Company Town 111
Merry Last Christmas, Jack Dorsey 116
The 2020 Guide For White Men In Tech 121
What If We Get Through This? 125

THE WORKING ANGERS
Can Design Change The World? 133
Yes, I Will Shame The Workers 136
Stop Covering Your Ass 140
Who Do Designers Really Work For 144

TABLE OF CONTENTS (CONTINUED)

Bad Work Is Always Your Fault...149
When To Put Down The Tools..153
10 Things You Need To Learn In Design School
If You're Tired Of Wasting Your Money....................................156
Getting Your First Design Job..161
Our Primary Contract...165
Starting A Studio..174
Eight Reasons To Turn Down That Startup Job......................176
Get Paid What You're Worth To Speak At Conferences.........180
Thirteen Ways Designers Screw Up Client Presentations.....186
How To Read An Email..192
Stop Adopting Other People's Anxiety....................................197
How To Pitch A Project...200
Get Paid!...204
Everybody Leaves...206

Thank You..211
About The Author..213

INTRODUCTION

I am writing. I am attempting to write. I want to have written. Mostly I want you to have read it. I want to put words down, send them to other human beings, have them read them, and nod in acknowledgement. It doesn't need to be a nod of agreement, or a nod of incited rage, or a nod that I've given you something to think about (you have enough to think about), just an acknowledgement that there is another human being out there. And that human being has communicated. A nod of proof.

I am still here—in the middle of a pandemic, in the middle of an attempted coup—and if you are reading this, so are you. For right now that is pretty fucking great.

I recently watched a documentary about the last days of Other Music, which used to be a wonderful little record store in NY. The scene that destroyed me was when the truck pulled up to the curb on the day they were clearing out the store. The record bins, shelves, crates, chairs, assorted furniture—they were all going away in that truck.

As he sees the truck, Josh, the owner, says "That's just a little bitty truck, we got a lot of shit to fit in it!" And you can hear the hint of terror in his voice. And eventually a tear in his eye. That store was a universe. And a universe is too big to fit in a little bitty truck. It has to be! And you realize that Josh isn't afraid that everything won't fit in the truck. He's afraid it will.

All the sweat, and labor, and panic that goes into making a universe has to be more difficult to haul away, doesn't it? Surely, it takes more than a little bitty truck to haul away all our shit. We have made so much shit! We spend our lives making shit, and buying shit, and collecting shit—be it things, or achievements, or relationships with other humans. We demand to take up space, be it physical, or emotional. We want to be in someone's heart. We want to be top of mind. Some of us write books to take up space on people's shelves and in libraries. Some of us make music to take up space in people's record collections. Some of us make art to take up space on people's walls. We paint murals in the street because we want our community to see us. We gather in the street and march to make our nation hear us. We build fences to point out the boundaries of our space to others. And when we die, some of us mark our space with a stone. We demand to take up space.

We demand space because space is memory. Once we no longer take up space we begin to be forgotten.

We want to plant a flag in the world's collective memory. We are so afraid that it will be easy to haul us all away when we are done.

It is very scary not to take up the space to prove our lives. It is very scary to be reminded everything we ever did, the good and the bad, the finished and unfinished, the celebrated and the regretted, can all be hauled away in a little bitty truck.

It is very scary to be reminded that we were not as important as we needed ourselves to be.

It is very scary how little it takes to make everything go away.

December, 2020. May this year go fuck itself twice.

PART ONE

The Righteous Angers

Gilly & Billy drawing by Adam Koford

> "The way you get a better world is, you don't put up with sub-standard anything."
>
> — Joe Strummer

MY PEOPLE WERE IN SHIPPING

(Originally published in Medium on November 30, 2020)

On Tuesday, June 16, 2015, Donald Trump descended on a golden escalator to the lobby of Trump Tower and announced that Mexicans were rapists. America laughed, because the whole thing was ridiculous. Well, some of America laughed. Some of America didn't. Some of America decided this was excellent. Some of America decided this was just the kind of message they needed to hear.

Trump was the "immigrants took our jobs" meme made pustulant flesh. Some of America, mostly white, mostly male, decided this was exactly what America needed. To grow up white and male, within a system that is designed specifically for you to succeed, and yet not succeed... Well, that's embarrassing, and Trump was giving those white males an out. They could blame immigrants. Which, let's be honest, a lot of those white males were already doing. But Trump was saying the quiet part out loud, and on November 8, 2016, some Americans, mostly white, mostly male, took the out Donald Trump was handing them, and made him president of the United States.

I am an immigrant. My parents, along with my brother and I, arrived in the United States on January 20, 1970. I was two years old. I wasn't born here, but this was the only home I'd ever known. So I joined the immigrant resistance (mostly behind the safety of a large shiny computer screen) and raised my voice along with the rest of my immigrant brothers and sisters in renouncing his xenophobic bullshit.

The first talk I wrote during the Trump administration was titled *How to Fight Fascism*. It ended with a slide that said MADE BY AN IMMIGRANT; a slide which I've copied over into every talk I've given since then. I was very proud to stand defiantly in front of that slide at the end of my talks. I convinced myself that I was standing in solidarity with other immigrants in the crowd, and I was, but there was a word missing. Yes, I am an immigrant, but my lineage is a little more complex than that. I'm a Portuguese immigrant. My people were in shipping. And as Letty in *Lovecraft Country* so succinctly put it—"that means slavery."

If you're African-American there is a very good chance that my ancestors and your ancestors crossed paths. And that your ancestors were free before meeting mine, but not after. Portugal invented the Atlantic slave trade. The Portuguese rounded up free people[1], imprisoned them in forts, and packed them in ships. Portuguese vessels carried an estimated 5.8 million Africans into slavery, mostly to Brazil, but also to the United States. The first Portuguese came to America in the stolen bellies of stolen African women. While historians, mostly Portuguese, will tell you that Portugal was a minor player by the time the Atlantic slave trade reached its zenith, it's kinda like saying a fire isn't your fault because the match you used to light it has gone out. These are my people. They were in shipping. They were slavers. And if we're going to make a case for intergenerational trauma, I believe it's not only fair, but necessary, to also make a case for intergenerational sin. We do not get a medal for solving a problem we had such a large hand in causing.

I am telling you these things because it makes me uncomfortable to tell you these things.

A BRIEF HISTORY OF IMMIGRANT RACISM IN AMERICA

In her excellent book *Caste: The Origins of Our Discontent*, Isabel Wilkerson makes a compelling argument that America is a caste system defined by color, and that color is an American invention. Before America, we were Fulani, we were Belgian, we were Kanuri, we were Irish, we were Kongo, we were Polish. America made us black and white, chained Black people to the bottom, and expanded and contracted the definition of white as needed to make sure Black people stayed on the bottom.

One of the surprises of the 2020 Presidential election was that Trump's percentage of immigrant votes grew. By this I mean that my white friends were surprised. I was not surprised. Let's talk about immigrant racism.

To look at me, I am white. I have certainly benefited from my

---

1 Because I have no problem shaming people who need to be shamed, here's a comment that was left by another white dude when I posted this on Medium: "The Portuguese didn't 'round up free people', they mostly just bought slaves off of African tribes who enslaved others as war booty." Imagine the flex and shame required to blame someone for their own enslavement. Sadly this theory of "It's not our fault, they did it to themselves" is fairly pervasive. We're terrible people and we have a lot of work to do.

skin color throughout my life, but that whiteness was a suit I had to learn to wear. When my family moved to Philadelphia in 1970, they were moving into one of the most racist cities in America at the time, presided over by racist mayor Frank Rizzo. We moved into a small Portuguese community in a majority-Black neighborhood. We moved into homes and businesses recently vacated by white flight. We came in as Portuguese, and we needed America to make us white, because that is how America defines success, and we were here for success. (The irony of having to find our place in a caste system we helped to create is a cursed monkey's paw implementation of John Rawls' veil of ignorance, but I'll save that argument for someone who didn't go to a state school.) We hung the Rizzo re-election signs in our storefronts, later we would hang the Reagan signs too. We crossed the street when Black people came our way. We hired our own. And we adopted all the slurs. Our goal was to achieve whiteness, which meant hating blackness and hating immigrants. Every immigrant group that comes into America wants to be the last group through the door. Trust me, we would rip the plaque off the Statue of Liberty faster than a Proud Boy at a tiki torch Black Friday sale. And every immigrant group knows the secret to achieving whiteness—patiently wait in the wings until the current whites believe Black people are catching up, at which point, the books are open, and the Irish are let in, or the Ukranians, or the Czech, or the Cubans. In America, whiteness is a reward for stepping on the necks of the underprivileged.

So, no, seeing that Trump had gotten more immigrant votes in 2020 than 2016 didn't surprise me. After all, when I attempted to talk to my own family about his xenophobia my mother's reply was "Oh, he doesn't mean us!" There's always an immigrant group on deck for achieving whiteness. They're voting for their turn at bat. And my family did indeed vote for him. But when you say that immigrants don't vote in their own self interest that's not true. We are voting in our own self-interest. We understand how this country was designed to work. We're playing by the rules you set. We did, indeed, learn it from you, Dad.

I am telling you these things because it makes us both uncomfortable when I tell you these things.

LET'S TALK ABOUT SHAME

I was recently having a conversation with a friend about how people we know, and believe to be good people, continue to work at places like Facebook, despite the overwhelming evidence that places like Facebook are, you know... bad places. We discussed the obvious suspects: a good salary, overwhelming student loan debt, fancy job perks, and all those things are true to some extent. But I believe the biggest reason is shame. Once you admit your involvement in something terrible you have to deal with your shame. I'm not even talking about admitting your involvement to others, I'm talking about admitting it to yourself. To admit you've spent years working on tools to dismantle democracy is a shameful thing. Especially if you've continued working on them long after the point where it was obvious what you were working on was complicit in dismantling democracy. The easiest way to keep that shame at bay is to not admit those things are bad. Which is one of the reasons companies distract you with things like good salaries and fancy job perks. They're shinier than the shame.

If you ever find yourself in Lisboa—and I encourage you to go, it is a lovely multicultural city now!—you may find yourself staring at one of its marvels, the magnificent monument to its sea-faring past: Os Descobrimentos. The monument points out over the Rio Tejo like a giant arrow, and it's adorned along the sides by action-posed statues of the great navigators of Portugal. My forefathers. This monument is a bauble. It is meant to take your mind off other things.

Just six kilometers to the east of that monument—and I encourage you to walk it because it's a nice walk—you'll walk into Pelourinho Velho (Old Pillory). It's a public square. It once served as Portugal's premier slave auction. Walk a kilometer to the northwest of that and you'll end up at Rua do Poço dos Negros (Street of the Negro Pit), where my ancestors threw the lifeless bodies they'd exhausted. There is no monument in either place. In fact, there is no monument, or museum, in Portugal dedicated to its slaving past.

We don't erect monuments to shame. In fact, our slaver past can best be summed up by this quote from Renato Epifânio, president of the International Lusophone Movement: "Anyone who knows anything about Europe has to agree that Portugal is probably the least racist country in Europe. This can, and should, be one of our greatest causes of pride." It can't, and it shouldn't be. The smallest asshole at the asshole party is still an asshole.

But let's get back to the United States... because I am telling you these things, and making myself uncomfortable, only to make myself feel better about how uncomfortable I am about to make you.

After all, we were slavers because there was a market for it, but the hand of the market isn't always invisible. It has fingerprints.

NOW I AM WHITE

For the first two decades of my life, being an immigrant was my defining characteristic. The neighbors saw me as an immigrant. The other kids at school saw me as an immigrant. The officer at the unemployment office where I'd go with my dad to serve as a translator saw me as an immigrant. And coming home crying after getting my ass kicked after school only to have my mother tell me, "this is their country, not yours," made sure that immigrant was etched deeply into my foundation.

Once I left the immigrant bubble to go to college I got to put on my white suit. Again, my mother sagely warned me that, "immigrants don't waste money on stupid things like art school." On a college campus, I was now surrounded by people who I could choose, or not choose, to reveal my immigrant-ness to. I'd achieved the dream, climbed the caste system and claimed my whiteness. Huzzah. And then, proved it wholeheartedly by painting anything I achieved as a product of hustle, hard work, and intellect. Double huzzah. It felt great for a while, until it didn't.

This was right around the time that I found out that although, yes indeed, I was an immigrant, the history of my own people was a little extra than most. It wasn't talked about at home, or in the community where I grew up. It was during a conversation with a Black classmate in college. We were talking about where we'd come from. I mentioned I was an immigrant. "No shit? Where from"? "Portugal." His face changed. His body language changed. I asked him what was up. "Oh, you don't know?" I didn't. He told me. We both avoided each other after that, unsure of how to handle it. I'm sure I didn't handle it well. That was shame.

As an immigrant, you get to be excited about America's future while taking a mulligan on its past. As a Portuguese immigrant, well... my people were in shipping. We are the foundation of America's past.

There is a line in Ijeoma Oluo's excellent book *So You Want to Talk About Race* that I've used in essays and talks before, and it bears repeating here: "If you are white in a racist society, you're a racist. If you are a man in a sexist society, you are sexist." By which she means, if I may mansplain, that people who look like me get these privileges regardless of whether we want or not. I'm quoting this line again because the first time I read it my reaction was that obviously this didn't apply to me. How could it? I'm so woke! I put up a slide saying I'm an immigrant at the end of all my talks! I am pulling out this quote one more time because of how deeply uncomfortable it made me when I first read it. It covered me in shame. Of course the line applied to me. The first time I read that line I spent the next twenty pages of the book attempting to read what she'd written, but it wasn't sinking in because I kept going back to that line. I couldn't hear what she had to say because all my energy was going towards keeping that shame her words had awoken in me from hitting home. I'm also ashamed to say how long it took, but once I accepted them I could actually hear what she had to say. I had to acknowledge and own my discomfort and shame. And that's when the work begins. Now is the time for people that look like me to be uncomfortable.

In twenty years of running our own studio we have hired exactly one Black person. Sentence about the pipeline yadda yadda yadda. Sentence about how you can only hire who applies yadda yadda yadda. Once all the excuses that you attempt to use to cover your shame are exhausted you're left with the truth: we have hired exactly one Black person in twenty years. That's a fact. And it's a fact that makes me very uncomfortable, because uncomfortable is where we need to be. It's also a fact that most of the companies in our industry are no better. You have not hired enough Black people. And if reading this is making you uncomfortable, great.

But Mike, isn't this a quota system. Maybe. But believing that every white dude that walks through the door is qualified by default is definitely a quota system, and it's the longest running quota system in world history.

YOU SAID BLACK LIVES MATTERED...

On March 13, 2020, Breonna Taylor was asleep in her bed when three Louisville violence workers busted into her house and murdered her in her own home. They got away with it.

On May 25, George Floyd was murdered by Minneapolis violence workers. He was handcuffed, thrown to the ground, and a violence worker lodged his left knee on George Floyd's neck for 8 minutes and 46 seconds. George Floyd used what breath he could muster to tell the surrounding violence workers that he couldn't breathe almost thirty times. How many times does someone need to tell you they are killing you? How many times did the surrounding violence workers hear it and ignore it? They got away with it.

This is America working as designed. The caste system that was designed not just by America, but in order to create America. White on top. Black on bottom. A caste system that was systematically enforced, first by slavery, then by Jim Crowe, the klan, redlining, restrictive covenants, police departments, every headline referring to the death of George Floyd rather than the murder of George Floyd, and finally our government itself, which is currently, as I write this, attempting a coup by throwing out votes in the majority-Black cities of Detroit, Atlanta, and Philadelphia. (The crime is in the attempt, not the success.) America is so addicted to its racism that most conversations, even and especially white liberals' conversations, about healing still revolve around pardoning Trump and finding common ground with his base, which is like treating a bullet wound by polishing the gun.

In the summer of 2020, part of America took to the streets. Spearheaded by Black Lives Matter, who in the middle of a pandemic, managed to organize millions of people safely so they were masked and socially distanced in cities and towns around the country. They were met by violence workers as well. In the wake of these protests, companies and organizations and corporations posted pronouncements in support of Black Lives Matter. Some promising donations, some promising to change their own cultures, some promising both. Some telling you how woke they were. Some letting you know they would've voted for Obama a third time. (Beware, some of them think they just did.) And while we could spend forever debating how many of those pronouncements were cynical (more than zero), and how many were genuine (less than a hundred percent), for our purposes the important part is that they did it, and they did it publicly. Which, in the business, is what we call a receipt.

Now is a great time to ask the people who made those pronouncements what their next step is. It is a great time to hold ourselves accountable. It will make us uncomfortable. Let it.

...NOW PROVE IT.

The future will bring white men with solutions. We cannot seem to help ourselves. It's what we do. Like colonizers with smallpox blankets we will show up looking to solve the problem we created, but to our own benefit. We hope to profit from the disease and the cure. Some of the white folks helping to pull down racist statues thought they were clearing space for statues of themselves. The truth is that I hope we are done idolizing individuals. Individuals will always let you down.

We do not get to speak for people whose ancestors we silenced. We get to listen. We get to stop hogging the space. We aren't giving anyone space, because the space was never ours to take and it is stolen. So if anything we are belatedly giving it back. We need to be uncomfortable. I am writing this essay in praise of discomfort because discomfort is where we need to be. For years, for decades, for centuries if you looked like me you got to live in a world that was very specifically designed to make you comfortable. (The fact that you, individually, didn't achieve it doesn't mean it didn't exist. It means you couldn't score from third on a double.)

You can turn your shame and discomfort into rage, as so many Trump voters did, or you can own it. You can claim it, because it's yours. And in so doing, you can keep from passing it on to the next generation. You can be a better ancestor than the ones you got.

There will be a role for people who look like me, but it'll be a role that we're very unaccustomed to: listening rather than speaking, giving before we take. It'll be as part of a community, and if we're lucky we'll have as much chance to succeed as anyone else in that community. And the community's welfare will be the most important barometer to success. If we're lucky people will treat us better than we've treated them.

It'll make us uncomfortable, and that'll be a good start.

ON MY SECOND BIRTHDAY WE LANDED ON THE MOON
A fifty year retrospective of trying to live with all of you on this goddamn planet

(Originally published in Medium on July 20, 2017)

On my second birthday, we landed on the moon.

Obviously, I have no actual memory of watching it happen. No memory of being huddled around a small black and white television in the family den. No memory of watching this happen while wearing a birthday hat, mouth lathered in the remnants of cake. I was two years old. I'm not even sure my family had a small black and white television, or a family den for that matter. Most importantly, when I say "we" landed on the moon I have no fucking idea who the we is.

When we landed on the moon, my family hadn't immigrated to the United States yet. We were still in Portugal. Living under fascist rule. And when my father saw a rocket escape the atmosphere and saw an American flag planted on a celestial body he could see in his own sky with his own eyes, I think he decided that flag was a pretty good indicator of where freedom came from. And he decided to emigrate. (I've never had the heart to tell him the rocket was developed by fascists.)

I'm writing this because today's my fiftieth birthday. And I didn't want to let it pass without some sort of accounting. And after fifty years of not dying, I've decided I can take the time for a little self-reflection. On my second birthday, we landed on the moon. And 48 years later, *we* is still the most difficult word in that sentence. And it's the word I've been trying to get an understanding of for 50 years.

When we (my family) arrived in the United States in 1970, we settled in Philadelphia because it was the home of a lot of Portuguese immigrants from the small town my parents (and I guess

I) came from. We (my family) became we (a community of immigrants who looked out for each other.)[1]

We shopped at a Portuguese grocery store because they gave us credit. We rented from a Portuguese landlord because he wasn't concerned about a rental history. And my parents worked for Portuguese businesses because we didn't come here to steal jobs, but to create them.

When there was trouble, we were there. When someone was laid off a construction job for the winter, we cooked and delivered meals. When someone's son ended up in jail, we found bail. And when someone's relative wanted to immigrate, we lined up jobs and moved money to the right bank accounts to prove solvency.

On my second birthday, we landed on the moon. And 48 years later *we* is still the most difficult word in that sentence.

As anyone who has ever grown up in an immigrant community knows, a *we* demands a *them*. They were not us. And they didn't think so either. At the risk of airing immigrant dirty laundry in public, I can attest that immigrant communities can be racist as fuck. We hated Blacks. We hated Puerto Ricans. (It wasn't too long ago I had to ask my mom to stop talking about "lazy Puerto Ricans" in front of her half-Puerto Rican grandchildren.) We hated Jews. In our eagerness to show Americans we belonged, we adopted their racism. (We also brought some of our own with us.)

We hung Ronald Reagan election posters in our stores, even as he worked to bleach America. And the more I saw that our definition of ourselves was being defined by our exclusion of others, the more uncomfortable I became. Oh, but trust me, I'm not painting myself as an exception to this. I did participate. I said shameful things. I thought shameful thoughts. I fought for shameful reasons. Every fight at school was between races. And as an immigrant, I was never white enough not to get my ass kicked by the Irish kids, and just white enough to get my ass kicked by the Black kids. But my mouth earned every one of those beatings.

The only *we* I saw was that we all hated each other. And to survive, I needed to keep looking for a better *we*—and although I was too young to realize it yet, a better me.

...

On my sixteenth birthday I came home with a broken nose, and my mother reminded me that we didn't live in our own country, and would never have the same rights as those who'd been

---

[1] We're also the worst immigrants, which is something I wasn't ready to deal with yet in this essay. I eventually began to deal with it in the essay you read just before this one, My People Were in Shipping, three years later.

born here. (Thirty some years later we would elect Donald Trump president, ultimately proving my mother right.) I wasn't willing to accept this. America was the only country I knew. I watched the same TV as American kids. I ate the same junk food. I listened to the same music. I watched the same sports. How was I not an American too? How was this not my country? And if not this one, then which? And yet, as we drove to the emergency room, my mother crying, my father in a rage, and me with a bag of frozen peas (we had adopted some American comforts) held up to my bloody nose, I had to admit that when other Americans looked at me, they didn't think *we*. They thought them.

Luckily, Reagan's America provided a solid solution for sixteen year olds living in Philadelphia who saw themselves as outcasts: punk rock! Turns out the combination of art school and punk rock was an amazing drug for a kid who felt like an outcast. Come to think of it, drugs were an amazing drug too!

The world didn't want to include me. But the world didn't want to include a lot of people. And for a brief moment in time we found each other. We met in abandoned halls, we met in the back of steak shops, we met in shitty bars, we met in basements. But we met. And we supported each other.

The beauty of mosh pits is that people only knocked you down so they could pick you back up.

It didn't matter that we couldn't play instruments very well. (I never even attempted.) We played them anyway, and if you played them faster, no one realized you messed up. It didn't matter that no one would publish our writing. We made our own zines. It didn't matter that no one made clothes we liked. We put our own together with safety pins. We didn't need you to let us into your community. We made our own. And our community was made up of all kinds of people. All backgrounds. All sexual orientations—with revolving doors to choose a new one whenever you liked. In these basements, in these halls, in these art school classrooms, I met the most amazing fucked-up people and they loved you for who you were. Here in these halls of kindness I met people who I'd previously avoided, been afraid of, mocked, reviled, and flat-out hated and they were calling me brother. And I loved them back.

We weren't afraid to do anything because we decided Reagan was going to kill us all. We were invincible because we decided we were already dead.

Meanwhile, underneath all this fun and excitement, and drugs, was depression. The escape velocity I used to break free of the gravity of my parents' house was exhausted. And the moon loomed over me like something that had to be dealt with at some point. My brain, which is shared by me and my depression, told me it was time to run.

I took off to see other parts of America. I needed to see who else was out there. I knew the moon would follow. It took me another ten years to deal with my depression.

...

On my thirtieth birthday, I held my infant daughter in my arms.

I had no template for being a good father. My father didn't either, so I'm not putting this at his feet as much as I'm acknowledging it. But I wanted to be a good father. And when I went back in time, I couldn't find a template for it. My father and I, now that was a troubled we. (The same could be said about me and my mother.) And as I looked at my new daughter, I knew that we were in unchartered waters. I would have to figure out how this relationship—this thing between father and daughter—could make room for happiness.

I realized that for the first time in my life I was in a relationship I could not run away from, could not put on someone else, could not half-ass, could not pretend to do right. Even if I managed to get all those things right, what genetic malfeasance had I saddled this kid with? I looked at this little bundle of pink flesh and spit and poop and realized that inside her there was the genetic code for depression, Alzheimer's, cancer, anxiety, and all sorts of other shit. I looked at that little kid and thought, little one you are fucked.

My daughter made me realize that we are also a genetic code that travels through time. We are ancestors. We are descendants. And while we can't fix the problems of the past in the present, we can make sure that we break the patterns that formed those problems. We can make sure the problems of our ancestors don't plague our descendants. I want to be a good ancestor. Because I want my daughter to be a good ancestor someday too.

We don't have to be the people our parents raised us to be. We need to be the people our children need us to be. The truth is our children raise us more than our parents ever did.

I think I've been a good father. I can say beyond a shadow of a doubt that she has been a good daughter. She has pushed me to do things I was previously afraid to do. Because I didn't want her to be afraid to do them. She has taught me kindness, she has taught me patience, and she has taught me how to love someone unconditionally. But most of all, she has taught me that we have to stick around and get things as right as possible.

...

In my forties, I got over my fear of public speaking. I found out that not only did I enjoy it, I was good at it. And so people would invite me all over the world to speak in front of them.

I found myself in places I never expected to be: overlooking a glacier in Norway, dancing in an anarchist bar in Copenhagen, bartering in a Muslim market in Hyderabad, bouncing in an inflatable castle full of sex workers in Melbourne, biking through Malmo, eating falafel in Paris, setting off (minor) explosives in Budapest, riding bullet trains in Japan, sneaking horribly flavored vodka into old Soviet buildings in Warsaw, drinking in lesbian or gay or oh-who-the-hell-cares bars in Berlin. And with each trip, I met more people, and with each trip my definition of *we* increased. With each trip, I realized that we had so much in common, and that we had so much that was weird and different and that both were just as amazing.

I went back home to Philadelphia this past January. The city where I grew up an immigrant, and got beat up for being an immigrant, and joined a pro-immigration march at the airport. We chanted together. We marched together. We resisted together.[2]

We don't live in a bubble. We live on a bubble. A fragile fucking bubble which withstands so many pricks. So many pricks. So many pricks shouting same same same same. So much fear of difference. So much energy spent hating what is different from the small definition of we that we grew up with. So much energy being wasted, that we get further and further away from the moon with every subsequent generation. But every time you reach across the world and call someone brother or call someone sister, you are kicking against those pricks.

Nine years ago, I was on a flight from Louisville, Kentucky. We were disembarking kind of slow. Annoyingly slow. When I finally got out to the gate I found out why. There was a small crowd gathering around one man. Getting their pictures taken with him. He was holding himself steady on a walker. His hand trembled as he reached out to shake people's hands. But his smile was unmistakable.

I'd shared a flight with Muhammad Ali. For three hours our paths on Earth crossed just close enough to put us on the same

---

2 On November 7, 2020 I got to see those same Philadelphians dancing in the streets after Pennsylvania was called for Joe Biden, putting him over the 270 electoral vote threshold to knock Donald Trump out of office. About a month earlier Trump had mentioned that "very bad things happen in Philadelphia." Fuck around and find out, Donald.

plane. For three hours we'd breathed the same air. I cried that night. Standing there looking at someone who'd done so much to expand who we think of as we. I cried again the day he died. And I remembered those three hours we shared the same air. And I realize that in a larger sense, from the day I was born until the day he died, we'd shared the same air while traveling on an even bigger ship traveling through space. And I considered myself lucky.

On my second birthday, the moon hung in the sky like a glowing pin on a map. That same moon had hung in the sky for all the generations of ancestors. But this time it was different. This time, we were on it. This time, we were looking back. For the first time in history a human being took a photo of the entire planet.[3]

We did the celestial version of running across the street to get a photo with all of us in the shot. And in that shot, if we look closely, we'll see a bunch of people trying to live both ordinary and extraordinary lives. We are bent over a table paying bills. We are watching TV. We are crying over a dead pet. We are preparing dinner. We are toppling dictators. We are filling out a school application. We are learning how to kiss. We are casting nets into the sea. We are arguing. We are making up. We are fucking. We are making babies. We are dreaming of faraway places we've seen in books. We are brave and nervous wrecks waiting in a dark living room for our father to come home so we can tell him our gender is not what he thought it was. We are telling our new daughter that they will *always* be loved. We are wondering if we'll ever have the courage to to talk to that girl or boy or non-binary human in our social studies class. We are looking up into the night sky deciding to take our families to a place where they might have a chance to be free. We are all in that picture.

...

In fifty years of sharing this planet with you, I've learned less than some and more than others. I've learned half what I need to and twice what I was ever expected to. But I feel pretty confident telling you this: life is messy. Sometimes glorious, sometimes sad, sometimes terrible, sometimes exciting, sometimes a garbage fire, sometimes there are moments you want to bottle up and keep safe so you can return to them forever and ever. And with all that, it is messy. But as long as our definition of *we* gets a little bit bigger every day, and our definition of *them* gets a little bit smaller every day—we have a chance.

---

4   It's famous. Google "big blue marble." That's the one.

DESIGN'S LOST GENERATION

(Originally published in Medium on February 18, 2018)

I was in the audience at a gathering of designers in San Francisco. There were four designers on stage, and two of them worked for me. I was there to support them. The topic of design responsibility came up, possibly brought up by one of my designers, since this was a topic we were talking a lot about at work. At one point in the discussion, I raised my hand and suggested, to this group of designers, that modern design problems were very complex. And we ought to need a license to solve them.

About half the room turned to me in unison and screamed NO.

As if I'd just suggested something absurd, such as borrowing ten million dollars to develop a smart salt shaker. (A device that actually received venture capital funding.)

"How many of you would go to an unlicensed doctor?" I asked. And the room got very quiet.

"How many of you would go to an unaccredited college?" I asked. And the room got even quieter.

(And before you weigh in on the condition of today's health care and education, which I grant are very problematic, let me just add that it's not the level of service with which we typically take umbrage, but rather our difficulty in accessing and then affording those services, which tend to be quite good.)

Turns out we like it when our professional services are licensed. In fact, if you've ever had occasion to use a lawyer, I'm sure you've been comforted knowing they've passed the bar. Their certification will be neatly framed right behind their head. Not to mention that even the cafeterias of Silicon Valley's most disruptive companies have to hang health department grade sheets where diners can see them. So, while you take a break from fighting against regulations that keep passenger vehicles safe, you can avail yourself of that burger which you know is safe thanks to the regulations inspired by the muckraking work of Upton Sinclair. A journalist. Or in libertarian parlance—the media.

Turns out we enjoy regulations. When they're in our interest.

This roomful of designers, however, was quite taken aback by the idea that our industry, an industry which now regularly designs devices which go inside human bodies and control our medication and write logic for putting driverless tractor-trailers on the street, should need professional licensing.

"Who'll decide who gets licenced," they asked.

I'm confident that if other professions have figured this out, we can figure this out. We can even look to their example. The last runner off the blocks can generally find their way by following the asses in front.

...

I had coffee with a colleague who teaches design at the local art school. (Why design is taught in art schools is worth another 10,000 words. I'll save it for later.)

"How goes it with the new crop of kids?" I ask him.

"Good! You know, they're surprising me. They're asking about things like sustainability, working in civic organizations, and ethics."

"This is new?"

"Yeah. Up until recently they wanted to know about startups, funding, and money."

"There's hope."

"There is."

And that's when I decided that we—and by we I mean those of us currently drawing paychecks for professional design services—are design's lost generation. We are the Family Ties era Michael J. Fox of the design lineage. Raised by hippies. Consumed by greed. Ruled by the hand of the market. And nourished by the last drops of sour milk from the withered old teat of capitalism gone rabid. Living where America ends—Silicon Valley.

We are slouching toward Sand Hill Road. We are slouching toward another round of funding. We are slouching toward market share. We are slouching toward entrepreneurship. And ultimately, we are slouching toward irrelevance. If we're lucky.

Because the longer we stick around, the more we're leaving for the next generation to clean up. And we've given them quite a bit of job security as it is. We are slouching because we were born without spines. When society desperately needed us to be born with them.

The Righteous Angers

The center did not hold

There are two words every designer needs to feel comfortable saying. No and why. Those words are the foundation of what we do. They're the foundation of building an ethical framework. If we cannot ask "why," we lose the ability to judge whether the work we're doing is ethical. If we cannot say "no," we lose the ability to stand and fight. We lose the ability to help shape the thing we're responsible for shaping.

Victor Papanek, who attempted to gift us spines in Design for the Real World, called designers gatekeepers. He reminded us of our power, our agency, and our responsibility. He reminded us that labor without counsel is not design. We have a skill set which people need in order to get things made, and that skill set includes an inquiring mind and a strong spine. We need to be more than a pair of hands. And we certainly can't become the hands of unethical men. A designer who loses their hands is still a designer, but a designer who doesn't offer their client counsel is not.

We are gatekeepers[1], and we vote on what makes it through the gate with our labor and our counsel. We are responsible for what makes it through that gate, and out into the world. What passes through carries our seal of approval. It carries our name. We are the defense against monsters. Sure, everyone remembers the monster, but they call it by his maker's name. And the worst of what we create will outlive us.

There's no longer room in Silicon Valley to ask why. Designers are tasked with moving fast and breaking things. How has become more important than why. How fast can we make this? How can we grab the most market share? How can we beat our competitors to market? (And for those of you thinking that I'm generalizing, and that your company is different, I am—and you may be. But you can't argue, even if you're truly different, there aren't days you feel

---

1 So here's a word that's certainly evolved over the years. My definition of gatekeepers comes directly (meaning I stole it) from Viktor Papanek's Design For the Real World. He uses it to mean that we're the people who keep bad shit from crossing through to the other side. It's intended in a positive sense. In the past few years I've seen gatekeeper used more and more as a negative, referring to people that block other people's access to opportunity. The word has either evolved, or we've started to actually listen to people we should've been listening to the whole time. Which, come to think of it, is usually how language evolves. Either way, language is a wonderful living, evolving thing, and fighting its evolution is pretty stupid. Evolution is how we grow. If I were writing this piece today I would have changed gatekeeper to a different word. I'm leaving it in because I wanted to acknowedge that yes, I used the word. Yes I understand and appreciate that its meaning has changed. And I'm all for it.

yourself swimming upstream.)

The current generation of designers have spent their careers learning how to work faster and faster and faster. And while there's certainly something to be said for speed, excessive speed tends to blur one's purpose. To get products through that gate before anyone noticed what they were and how foul they smelled. Because we broke some things. It's one thing to break a database, but when that database holds the keys to interpersonal relationships, the database isn't the only thing that breaks.

Along with speed, we've had to deal with the amphetamine of scale. Everything needs to be faster and bigger. How big can it get? How far can it go? A million dollars isn't cool. You know what's cool? You know the rest of the line. When we move fast and break things—and those things get bigger and bigger—the rubble falls everywhere.

Facebook claims to have two billion users. (What percentage of those users are Russian bots is currently up for debate.) But 1% of two billion is twenty million. When you're moving fast and breaking things (this is Facebook's internal motto, by the way), 1% is well within the acceptable breaking point for rolling out new work. Yet it contains twenty million people. They have names. They have faces. Technology companies call these people edge cases, because they live at the margins. They are, by definition, the marginalized.

Let me introduce you to one of them:

Robyn Mansions[2] was "accidentally" outed by Facebook when she was a college freshman. When Robin got to college, she joined a queer organization with a Facebook group page. When the chorus director added her to the group, a notification that she'd joined The Queer Chorus at Big State University was added to her feed. Where her parents saw it. Robyn had very meticulously made her way through Facebook's byzantine privacy settings to make sure nothing about her sexuality was visible to her parents. But unbeknownst to her (and the vast majority of their users), Facebook, which moves fast, had made a decision that group privacy settings should override personal privacy settings. Robyn was disowned by her parents and later attempted suicide. The designers broke things.

A year later I gave a talk at Facebook. I told Robyn's sto-

---

2 I've changed the name here. This story took place in 2012 and I've now told it in a talk, a book, and several essays. This woman has more than earned the right to move on with her life, and the point can be made without her actual name. Yes, you could google the original version of this essay. But that kinda defeats the spirit of the thing. Thanks.

ry, which was public at that point. An engineer in the audience screamed out "it was the chorus director's fault, not ours." And that somehow managed to be the scariest part of this whole story. We're putting the people who need us most at risk, and we're not seeing our responsibility. And to this I must both ask why and say no.

We're killing people. And the only no I hear from the design community is about the need for licensing. If why and no are not at the center of who we are, and they must be, the center has not held.

We need to Slow. The. Fuck. Down.

And to pay attention to what we're actually designing. We're releasing new things into the world faster than Trump is causing scandals.

WHY WE FAILED—THE FIRST REASON

"I want to do the right thing, but I'm afraid I'll lose my job."

"Must be nice to afford to take a stand."[3]

"I have rent to pay."

"If you're telling people how to work, then you're the fascist!"

I've heard variations on all of these phrases thrown at me from designers I've spoken to all over the world. Sometimes they're apologetic about it. Sometimes they're angry. Sometimes they're looking for absolution, which I'm not in a position to give. But mostly they feel tired and beaten down.

Yes. You will sometimes lose your job for doing the right thing. But the question I want you to ask yourself is why you're open to doing the wrong thing to keep your job? Without resorting to the level of comparing you to guards at Japanese internment camps, I'd argue there are paychecks not worth earning. An ethical framework needs to be independent of pay scale. If it's wrong to build databases for keeping track of immigrants at $12 an hour, it's still wrong to build them at $200 an hour, or however much Palantir pays their employees. Money doesn't make wrong right. A gilded cage is still a cage.

You'll have many jobs in your life. The fear of losing a job is a self-fulfilling prophecy. Fear makes it less likely that you'll question and challenge the things you need to question and challenge. Which means you're not doing your job anyway.

The first part of doing this job right is wanting to do it right.

---

3 *For the record, it's not that I can afford to do this. It's that growing up as an immigrant, I've seen a little bit of the effects of being marginalized. Certainly not as much as others, but generally more than the white boys (they're always white boys) who say this to me.*

And the lost generation of designers doesn't want to do it right. They found themselves standing before a gate, and rather than seeing themselves as gatekeepers, they decided they were bellhops.

We failed because we looked at our paychecks, saw Mark Zuckerberg's signature, and forgot the person we actually worked for was Robyn Mansions.

WHY WE FAILED—THE SECOND REASON

I also hear from plenty of people who attempted to do the job right and hit their head against the wall time after time—maybe it makes you feel better to put yourself in this second group. These were the people who looked for backup and didn't find it. Whether it was backup from within their organization, or the backup of a professional service who protects the integrity of the craft.

Let me tell you a story. My family and I drove out to Sequoia National Park a few years ago. We stopped at a diner on the way. There was an elderly couple sitting next to us. He was wearing one of those navy caps with a picture of the battleship he'd served on. When their check arrived, the old man wasn't happy with the total. He called the waitress over and informed her the check didn't reflect the early bird price. She smiled, and in her best voice, told him they'd been seated just a few minutes too late to get the early bird price. At which point Joe, and I know his name was Joe because his wife was telling Joe not to make a scene, reached for his wallet, pulled out his AARP card and sat it on the check. That was the end of the argument. No one fucks with the AARP. Because the AARP looks out for their old people, and they will fuck your shit up. Had that waitress not given Joe the early bird price, I'm pretty sure a platoon of AARP lawyers would've parachuted into the diner. Joe ended up paying the early bird price because Joe had the power of a professional organization behind him.

Imagine this same situation playing out with a designer standing up for the solidity of their work. Imagine the power of a professional organization having our backs. We've never had that. Possibly the AIGA came closest, but closest isn't even the right word because it contains the word close. They look at UX designers the same way Donald Trump looks at a vegetable. I do believe they had an opportunity to become the organization we needed, had they wanted to be it, and had they taken the time between poster contests to do some actual work.

But every designer out there fighting the good fight is doing it with the knowledge that they're going at it alone.

WHY WE FAILED—THE THIRD REASON

The history of UX design is, until very very very recently, the history of design as defined by other fields. Our field was defined first by engineers because, let's be fair, they're the ones who invented the internet. And their definition of design—the people in the bunny hats who make the colors—is still widely accepted by a large majority of designers working in the field today.

It's the easy path. You sit in the corner, listening to The War On Drugs on your big expensive DJ headphones, picking colors and collecting checks. We've spent the last twenty years proving our legitimacy to engineers who thought we were a waste of time. Until they realized we could magnify their power exponentially.

We let other people define the job. We complained when we were told what to do. We complained when we weren't told what to do. We became proficient in eye-rolling. (Be honest. You proved my point by rolling your eyes at that last sentence.) We fought for a seat at the table, and once we started getting that seat, we found out a lot of designers didn't want it.

I'm a little unfair when I say that designers haven't fought. We've fought to have other people define us. We've fought to have other people define our responsibilities. We've fought to give away our agency. And we've fought not to have a seat at the table. We were all too happy to dribbble away our time while decisions were being made around us. (Shade.)

A few months ago, Jared Spool, who's been doing yeoman's work for design for the better part of forty years, tweeted this out:

>
> **Jared Spool**
> @jmspool
>
> Anyone who influences what the design becomes is the designer.
>
> This includes developers, PMs, even corporate legal. All are the designers.
>
> 7:07 AM · Mar 1, 2017 · TweetDeck
>
> **648** Retweets and comments **923** Likes

Everything in that tweet is correct. Everyone who influences the final thing, be it a product or a service, is designing. And yet, if you click through and look at the replies, what you'll see is the evisceration of Jared Spool in defensive bite-sized vitriolic thoughts still covered with the spittle of ego. And, even more sadly, it quickly turns into a discussion of titles. We are happy to give away all the responsibilities that come with the job, but don't take our titles. I have seen designers argue for a week with a new employer about what their title will be, without sparing one breath to ask about their responsibilities.

Design is a verb. An act. Anyone is free to pick up the ball and run with it. And if you're not doing the job you're being paid to do, you can't be upset when someone else starts doing it. You cannot not design. What a professional designer brings to the act is intention. But for that, the designer needs to behave intentionally. Designers are dead. Long live design.

You just rolled your eyes. You should've thrown an elbow.

We are the children of capitalism's last gasp

We are all working in a system which measures success financially. We are about how much money a movie makes on opening weekend. We pay attention to music climbing up the charts, and Jack Dorsey's leadership was finally vindicated when Twitter posted their first profit-making quarter.

This was the first sentence of a Mashable article announcing Twitter's historic profit-making quarter: "It turns out cutting back, focusing, and maybe a little Donald Trump can help make money."

Let's look at the price of that profit-making quarter. Twitter's profit came at the cost of democracy. When an American autocrat chose it as his platform of choice to sow hate, disparage women and minorities, and dogwhistle his racist base, Twitter rallied. Rather than shut him down for violating their terms of service, Twitter chose to expand those terms of service to accommodate the engagement Trump was bringing them. Twitter, and every employee working within Twitter, failed their moment. Their ethics failed them. The reason Donald Trump has access to nuclear weapons is, in no small part, thanks to Twitter.

And yet, all is forgiven because they've now turned a profit. Profit justifies everything. Silicon Valley, the engine that powers the end of America needs to profit in order to survive, and it needs profit at scale. We remain enamored with our ideas, and blind to their effects. We award golden parachutes for failing big, because Silicon Valley rewards failing big over succeeding small.

The biggest sin in Silicon Valley is a small victory.

THE PATH FORWARD

In August of 2017, James Liang, an engineer for Volkswagen, was sentenced to 40 months in jail. A court in Detroit, Michigan sentenced him for knowingly designing software that cheated federal emissions tests. He wasn't the only Volkswagen employee sent to jail for this. (Thankfully.) But he's the one important to our story. He knew he was designing something deceitful, and he did it anyway. That's an ethical breakdown.

In March of 2017, Mike Isaac published an exposé in the New York Times about Greyball, a tool at Uber designed to purposely deceive authorities. Authorities that were looking out for the safety of Uber's riders. No one at Uber has yet to go to jail. But the stories are the same.

Two companies, both of which knowingly designed software with the express purpose of deceiving regulating bodies. Volkswagen got caught because the automotive industry is regulated. We know cars are dangerous. Uber got away with it because they claim to be a software company, (Narrator: they're not.) and we're just beginning to realize how dangerous software can be, especially in the hands of companies led by ethically bankrupt men. But Travis Kalanick, Uber CEO when Greyball was designed, should be in jail as well.

We need to be held accountable for our actions. We've been moving fast. We've been breaking things. Sometimes on purpose. Sometimes out of ignorance. The effects are the same. The things we're building are bigger than they used to be, and have more reach. The moment to slow down is here. Because what we're breaking is too important and too precious. Much of it is irreplaceable.

I am part of design's last generation. I've fucked up. We all have. None of us did enough. Maybe the tide was too strong, or maybe we were too weak. But as I look behind me, I see the hope of a new generation. They're asking better questions—at a younger age—than we ever did. And I truly hope they do better than us because the stakes have never been higher.

For this younger generation to succeed, they're going to need the following:

A BODY OF OVERSIGHT

You're going to need someone to have your back. Look to history. Long hours and working weekends are still long hours and working weekends, regardless of whether the cafeteria is serving swordfish. Human resource departments do not work for you, as Susan Fowler and many courageous others have found out. They work for your boss.

I am the son of a Philadelphia construction worker. Every winter my father got laid off because it was too cold to build, and every winter someone from the union showed up with groceries. The only ones who will ever stand up for workers are other workers. Not only do you need licenses, you need a union. When one of you is trying to do the right thing, let them (and their bosses) know that there's an entire brotherhood and sisterhood standing behind them.

EDUCATIONAL AUTONOMY

Art has as much in common with design as a potato has with a Honda Civic. Why are we still cramming design departments into art schools? Not to disparage art schools—they're a wonderful place to get an art education. I'm also not disparaging existing design programs here as much as I'm trying to get you more room! Design is too important and too big a field now to be given a wing in someone else's school. It's time to create our own. A few years ago Jared Spool and Leslie Jensen-Inman did just that. They started Center Centre, a small school in Chattanooga, Tennessee to specifically train UX designers. I hope it succeeds and that it's the first of many.

LICENSE TO PRACTICE

My friend Ryan is a professional dogwalker. Dude loves dogs. Which I totally get because at our best, our goal should be to be the people our dogs think we are. Before Ryan could become a professional dogwalker, he had to get a license. He had to pass a test. As someone who loves my dog more than I should, I'm glad he had to do that. It reassures me that my dog is in good hands. I know that if my dog does something stupid, which he does plenty, Ryan will know how to handle it.

My dentist is licensed. My doctor is licensed. My lawyer is licensed. My accountant is licensed. Almost every professional I interact with is licensed. There are really good reasons for that. Not only does this let me know they've passed some sort of test, some sort of proficiency, but it also gives me a way to measure a standard of expectation for their level of service, and a way to address any grievance with a lack of it.

Closer to home, architecture is the hardest design profession to study and to get into and has incredibly high standards. Architects can debate style and aesthetics all night, but at the end of the day, their shit has to be up to code. Architects have to make sure that engineers and contractors carry out their intention and are ready to make changes to their vision to accommodate reality. There's no minimal viable product in architecture because bad architecture kills people. Bad UX is now just as deadly. And yeah, even Howard fucking Roarke had to be licensed.

And while I'll be the first to agree that licensing doesn't solve all of the problems listed in this article, I do believe that it's the first step in addressing those problems. It gives us a chance. Let's not spend the next ten years looking for the perfect solution at the cost of implementing a good one.

...

As professionals in the design field, a field becoming more complex by the day, it's time that we aim for a professional level of accountability. In the end, a profession doesn't decide to license itself. It happens when a regulatory body decides we've been reckless and unable to regulate ourselves. This isn't for our sake. It's for the sake of the people whose lives we come in contact with. We moved too fast and broke too many things. Amateur hour is over.

HOPE IS NOT ENOUGH

(Originally published in Modus on November 27, 2019)

> *"We've been behaving so badly that I hope the government comes in and regulates us."*
> —anonymous Facebook employee I spoke with

The first car I remember my father driving was a 1973 Plymouth Gran Fury. It was midnight blue. Big as a boat. You could easily fit four people across the back seat. It smelled like cigarettes. He bought it used. But he looked good in it, which is probably why he bought it. My father was a vain man. Is a vain man.

The car got terrible gas mileage, which hadn't yet become a concern for most people. At least in a big-picture, fossil-fuel-emissions kind of way. We were still a little sore about losing the Vietnam War and needed something to remind us what it meant to be American. Large cars did the trick then. (Cybertrucks do it now.) The shitty gas mileage did matter in a small-picture way, though. My father was a construction worker in Philadelphia, where the ground freezes over in the winter. Construction stops. So do paychecks. And it took a lot of gas to fill that car's tank.

I remember one particular winter morning. We'd been hit by a snowstorm a few days earlier and the streets were still covered with snow. The city of Philadelphia didn't exactly rush to plow our neighborhood. He decided to drive us to school. We piled in the car. Got yelled at for not kicking the snow off our shoes before climbing in.

"We need to stop for gas."

We pulled up to the pump at the AM/PM minimart and my dad pulled out his wallet. Five ones. Went through the change compartment and dug out a few quarters and some more odd change.

"Be right back."

He came back in the car holding a lottery ticket and tucked it into a crack in the dashboard foam that he used for important paperwork, and we drove off to school.

"Hopefully we hit the numbers tonight."

My father lived on hope.

He never had a 401k. Never saved. Spent most of his life without insurance of any sort. Never planned much. But he had a lot of hope in outside forces. My father's life story was full of magical realism, where utility bill money showed up at the last minute, a loan could always be gotten, my mother could always find another reason to forgive him, and someone else would show up to take care of shit he was responsible for.

(Let me take a minute here to celebrate the eldest children of immigrants who might be reading this. The ones who learned how to file income tax extensions at the age of five. The ones who were pulled out of school to do translation work at the social security office. I see you.)

Hope is believing other people will act in our behalf. Hope is a reliance on deus ex machina. Hope is keeping us from dealing with our own shit. That's not hope, that's responsibility deferred.

So yes, you can hope that the government comes in at some point and regulates your job. And to be clear, they should. And they will. Eventually. But waiting for them to solve the problem is responsibility deferred.

Ultimately, hope is giving ourselves permission to do nothing.

Regulation, if done right, will take years. We don't have years to sort out our mess. Additionally, the current administration is using regulation as a threat. Mark Zuckerberg has had two secret meetings with Donald Trump (that we know of) where regulation is being used as a stick to keep Facebook from vetting political ads. Facebook's decision to not fact-check political ads came after that first secret meeting. We don't know the aftermath of the second meeting yet, but we know they were joined by Peter Thiel, the CEO of Palantir, the company that builds immigrant-tracking databases for ICE.

So yes, regulation is the answer. Eventually. But asking for it now is like putting the fox in charge of the chicken coop.

LIVE WITHOUT HOPE

I grew up in the 1980s with the fear of nuclear war hanging over my head. Today's kids are growing up in a world that's becoming hostile to human life.[1] Both these threats were caused by humans. We had nuclear attack drills in school. Today, children have to endure active shooter drills. Both drills are just as effective.

As angry punk teens, my generation dealt with the anxiety of nuclear annihilation by deciding we were already dead. We exiled ourselves from society. We focused our anger into music, zines, comics, getting high, and watching Pee-Wee's Playhouse[2]. There was power in living without hope. There was power in turning fear of something we had no control over into anger that we could vent at will.

You can either hope you don't get in trouble, or not give a fuck if you get in trouble. Hope isn't going to solve our problems. But not giving a fuck just might.

YOUR INDIVIDUALITY IS EXHAUSTING

The reasons we hope someone else is coming to solve our problems are worth going into.

For one, the problems seem insurmountable. Take Facebook, for example: No one has ever built a social network that connects 2.5 billion people before. Much less a social network that profits off the data harvested from those 2.5 billion people. Much less a social network that is happy to profit from deceiving 2.5 billion people. It's uncharted fuckery. And finding yourself inside that situation has to be daunting. And isolating.

There's a reason they call you individual contributors. It's to make you feel alone. Except you're not alone. You're part of a giant workforce. And you also didn't find yourself there; you actively chose to work there. Which brings with it a set of ethical responsibilities. If you're hoping someone else solves your problem, it means you're aware the problem exists. And as the person hired to build the machine, it's on you to correct the machine's actions when it goes off the rails.

Look around at the people working with you. They all chose to work where they're working. Everyone there made a decision to take that job. And when you take a job, you have to do it right. The odds that you're the only person who's hoping for change are pretty slim. Start talking to each other. Communicate. Work together. You are not individual contributors. You're a workforce. And a united workforce can accomplish things individuals cannot.

---

1 *When I wrote this I was referring to the climate crisis. A few months later the Covid-19 crisis hit. They're not unrelated. So yeah, I hit this nail harder on the head than I thought I had. Every day our descendents don't just outright murder us is a gift we don't deserve.*

2 *We loved Pee-Wee so much our parents decided they needed to destroy him.*

Secondly, the repercussions are real. If you take a stand against a company, they have a million ways to fire you. That's real. And as much as I don't want you to lose your job, I'm not willing to sign off on you doing it wrong so that you don't lose it. If you're willing to lie to people so you don't get fired, you don't deserve that job, and you certainly shouldn't be anywhere near a machine that can do that much damage.

I've seen too many instances of designers at big companies complaining about being unfairly maligned and shamed for "just doing their jobs." Apparently when you face consequences for deceiving people, it hurts your feelings. If you're a designer, the work can't be about your feelings, and frankly, your feelings are exhausting. As is the imposter syndrome that drives you to look for validation. You earned the job the day you were hired. You earned the paycheck, but you also earned the responsibility that comes with it. Now do it.

You don't work for the people who sign your checks. You work for the people who use the products of your labor. If I were to put my hope in one thing, it's that you understand the importance of this. Your job is to look out for the people your work is affecting. That is a responsibility we cannot defer.

THE GOOD HOPE

I started this essay with a story about my father, and I didn't paint him in the best light. My relationship with my father is complicated, but it's been considerably smoothed over by both time and therapy. For all the questionable decisions he made in his life (and I've made plenty of my own), this is the same man who got on a plane with my mom, two infants, and $200 in his pocket because he hoped America would be a better place to raise those infants. But that was hope combined with a plan and a lack of fear to act. It took the combination of all three. Hope by itself would've meant doing nothing. It's the difference between hoping for the best, and hoping our plan works. So I can vouch that at least once, my father coupled hope with a plan. And I'm grateful to him for that.

When 22,000 Google employees staged a walkout, it gave me hope. When Microsoft workers successfully protested their company's ICE contract, it gave me hope. Hope is earned by action. When you stand up against bad practices in your workplace, you earn hope. When you act to organize your workplace, you earn hope. When you're willing to risk your own comfort for the dignity of others, you earn hope. When you take action, you earn hope.

But hoping that people who are clearly benefiting from the broken state of the world will change their behavior is lost hope.

Hoping that enough crumbs will fall from billionaires' tables is lost hope. Hoping that others will do what we don't have the courage, resolve, and confidence to do is lost hope. More importantly, it's responsibility deferred. This is the job we signed up to do.

We need a plan. Then we need to act. Then hope is earned. Until then, it's just keeping us from acting.

ETHICS OFFSETS ARE BULLSHIT

(Originally published in Modus on August 21, 2019)

> Q: I work at a large social media company that does some shady shit. The team I'm on doesn't do the actual shady shit. We're shit-adjacent. I'm actually proud of what my team does. And I like working there. So every once in a while I'll donate part of my paycheck to a good cause, or a candidate who wants to regulate some of the shady shit that we do. Does that make up for the shady shit?[1]

Great question. The answer is no, it does not make up for the shady shit.

First off, let me applaud you for the decent things you do, because they are in fact decent. It's a damn nice thing to donate your money or your time to an organization or a cause that needs it. And I certainly don't want to talk you out of that. Please continue, and urge others to do the same.

But that's not what you asked me. You asked me if it made up for the shady shit. Which means you're looking for an ethical offset. In other words, Does the good I do over here make up for the bad I do over there? The idea of ethical offsets has been around for a long time, even when they weren't called that.

Let's get religious. For hundreds of years, the Catholic Church recognized something called "indulgences" as an actual part of how they did business. (It wasn't until the Second Vatican Coun-

---

1   So yeah, both Dear Design Student and Dear Designer were advice columns and I asked people to send in questions. Sometimes they did. Sometimes the questions were so specific to the writer's situation that I'd end up just emailing them a response and seeing if I could extrapolate something more universal from their question. And sometimes I just had a head of steam about a topic I wanted to write about and reverse engineered a question out of it later.

cil of 1959 that they suggested reforming this. They also did away with saying the Mass in Latin and meatless Fridays at the Second Vatican Council, commonly referred to as Vatican II, like a Super Bowl. It was a good meeting. There are notes.) According to the Vatican, an indulgence was a way to reduce the amount of punishment one had to undergo for sins. The original idea was that if you did a nice thing, the pope would issue you an indulgence which you would somehow present when you got to heaven. As a character reference. And the people at the door would weigh it against the shitty things you did. It was a reward. In reality, indulgences became a commodity which were bought and sold in the dark musty corners of cathedrals to shady characters who wanted to be dicks and treat other people like shit, but were still worried about getting to heaven. Like, if you wanted to kill someone you could get an indulgence ahead of time, and then you wouldn't have to feel bad about it. Theoretically, you could also get the indulgence after the murdering, but you'll probably end up paying more because you've already committed the murder and now it's a seller's market.

Here are two fun facts about indulgences. First, most people think the first thing on Gutenberg's press was the Bible. That's bullshit. It was indulgences. Almost no one could afford a Bible, but indulgences sold like hotcakes. Gutenberg was a good businessman. Second, indulgences were one of the main reasons Martin Luther went ballistic against the pope.

Because indulgences were a way to pay off the shitty behavior you were committing, they were expensive. Which meant the richer you were, the shittier you could behave. You could just buy your way out of it. Not unlike how the justice system works in America today.

If ethical offsets were a wrestling belt, The Catholic Church would be our first champion. And they held that belt for a long long time. Until they finally passed it on to the business world.

The percentage offset

Obviously, the idea that something good can offset something bad is a long-held belief that has propped up many a jewelry store and flower shop.

In *Design for the Real World*, Victor Papanek calls for designers to donate 10% of their time to non-profit causes and good works. On this, and maybe only this, Victor and I disagree. But Victor wrote that in 1971. Advertising agencies, who made a ton of money selling cigarettes and were looking like massive assholes after the 1964 Surgeon General's report that ended the nicotine-fueled advertising gold rush, needed a way to claw back from the public's hate. Offsets were in the air. Advertising agencies began carving away a percentage of their time for pro-bono work.

And so the Ethical Offset Belt moved to the advertising agency, which in turn begat the design agency, and the pro-bono percentage offset came with it, eventually passing to the startups of Silicon Valley.

There's a rich corporate tradition in Silicon Valley of granting employees a percentage of their time to "do good."[2] This is a nice distraction from what they're doing the rest of the time. Google is famously known for their practice of "20% time," when employees can focus on side projects. But as former Google HR VP Laszlo Bock wrote in his book Work Rules, not a lot of employees actually use it, which doesn't matter as long as the idea of it exists. It's guilt-relief as an interface. An indulgence.

The problem with the percentage offset is compartmentalization. On Friday I think about doing good things. The rest of the week I'm cool. And if I do something that leaves a bad taste in my mouth on Wednesday, I'll make up for it on Friday. Even if I don't actually end up doing it on Friday, the idea that I could if I wanted to, or wasn't too tired, makes me feel better about myself.

There's no percentage of your time that can make up for building databases at Palantir that ICE uses to round up immigrants. There's no percentage of your time that can make up for building facial ID software, also for ICE, while working at Microsoft. (In fact, those employees rebelled and forced Microsoft to cancel that contract.) There's no percentage of your time that can make up for allowing white supremacists to run free on Twitter. I don't believe in ethical offsets. There's no way that saving 10% of the world while destroying 90% of it turns into anything close to a net positive.

THE PASSION PROJECT OFFSET

Another popular way to deal with the guilt and shame that comes from doing unethical work is the much-lauded personal project. Again, this takes an amazing amount of compartmentalization. A stress-inducing amount. It practically forces you to live two separate lives. If you spend your day helping Yelp swap out restaurant's actual phone numbers with Grubhub-affiliated ones and then go home to work on a blog celebrating the local food scene, you still spent your day helping Yelp swap out that restaurant's phone number with one that wasn't theirs. So no, the local restaurant scene doesn't owe you a medal.

---

2   *It's also a great way for companies to take ownership of an employee's ideas.*

Franz Kafka spent his days working at an insurance company, and his evenings working on his passion projects. He wrote about turning into a roach. I don't think Kafka successfully compartmentalized. You can't either.

THE DONATION OFFSET

You should donate to ProPublica. You should donate to The Trevor Project. You should donate to RAICES. You should donate to Planned Parenthood.[3] You should donate to a ton of other organizations that need money to keep doing good work. You should donate locally, nationally, and internationally. You should donate to political campaigns. You should donate to the organizations that I hope people will mention in the comments to this piece. You should donate as much as you can, and the more you earn the more you should be donating. And if you cannot donate money, there are organizations where you can donate your time. Some of us may even be lucky enough to donate both.

That said, no amount of money, or time, that you donate to RAICES will make up for the fact that you earned that money while working at Palantir building a database ICE uses to separate children and parents. If you really want to help immigrants, and I hope you do, destroy the database.

STAND IN THE PLACE WHERE YOU WORK

My purpose in writing this isn't to make you feel bad. You obviously already do. As my grandfather once told me, you don't need to push a stone that's rolling downhill. My purpose in writing this is to give you good news. And here it is. If you are looking to make a difference in the world, you are already in the place you need to be. You don't need to go anywhere else.

I recently ran across a few tweets of Twitter employees from various offices cleaning up local beaches and parks. They were all tagged #lovewhereyouwork. I'm glad they love where they work. And I have good news for them. The next time they're looking to clean up trash they don't even have to leave the office. I am happy to go clean up the beach. I bet I can even get a big group to come with me. But only you can clean up the place where you work, and if you want to take a stand, if you want to make a difference, it needs to start at the place where you draw your paycheck. Because if you are earning a living somewhere that makes the world a worse place, there is absolutely nothing more important than you can do than to take a stand right there.

---

3 *I'm an asshole for leaving Black Lives Matter off this list. Please donate at blacklivesmatter.com*

Yes, the non-profits need your help. Yes, the campaigns can use your money. Yes, you should go canvas for a candidate. Yes, you should go out and clean beaches and protest inequality.

But all of these things need to happen in addition to doing your job well, and doing it in society's best interest. Not instead of.

Ethical offsets are bullshit.

ETHICS AND PAYING RENT

(Originally published in Dear Design Student on March 29, 2017)

Inevitably, when I bring up the topic of designers working ethically, someone will reply with some flavor of "that's nice, but I have rent to pay." Feel free to substitute "rent" for student loans, childcare, medical costs or any other very real and very valid concern. Along with what I'm guessing is a not-insignificant amount of designers who are filling in that blank with [lifestyle to which I've grown accustomed.]

Let's deal with the first group, since I have close to zero fucks to give for the second group. Either way, this promises to be less than an enjoyable article for both groups. It will neither give you permission to work unethically, nor outline a set of situations where working unethically is acceptable. If you think those reasons exist, you'd be wrong.

THE FALLACY OF SUCCESS

Where does this idea that you have to be open to tossing your ethics out the window to be successful come from? That's worth exploring a bit. Certainly, if we look around at the current landscape, we'll find plenty of examples of people who behaved—or continue to behave—unethically and have done very well for themselves. From Travis Kalanick to Donald Trump, we see people who've broken the rules (pardon me—disrupted!), skirted regulation, and have generally behaved abominably toward others, to much success. In fact, it could be reasonably argued that in those particular cases, their success is due to their lack of ethics.

But if we look closely at those same individuals, we also see the price they've paid for their unethical success—the lack of trust, the constant vigilance, the scrutiny, and the eventual comeuppance. History won't remember these people kindly, and for that matter, the present isn't viewing them very kindly either.

Are they successful? Yes. For now. And it's that little for now that you have to add that should give you pause. Their success is a house built on sand. Can you be successful by throwing ethics out the window? Yes you can. You can also eat three burritos in one sitting. But in both situations, that act is coming back up on you and it won't be pretty.

The fallacy of the road to success being paved by unethical work is just that, a fallacy. It's not a road. It's a dead end alley. It may provide a safe haven from the elements for a few minutes, but going from alley to alley to alley, hoping you don't get stopped, is a horrible way to complete a journey.

THE SLIPPERY SLOPE

We've all been in spots where we've done things that were ethically questionable. We're human beings. We're messy. For example, I think we can all agree that stealing is wrong. Yet none of us would hesitate to steal the proverbial loaf of bread to keep our families from starving. The problem comes when theft goes from being an emergency method to stave off starvation to the primary means through which you earn your income.

Throughout your career, you'll find yourself in spots where your only options might be doing a little work for one of the Travis Kalanicks of the world, or starving. By all means, don't starve! Just be honest with yourself about what you're doing, why you're doing it, and for how long you're going to do it. Because once you lose sight of that, the justifications start. ("I'm going to change things from the inside.") And realize if you keep making those decisions, they end up defining your career. The idea that you can work unethically, build up a reputation, and then swing that ship around into ethical waters is also a fallacy. By that point, you do indeed have a reputation, but not the one you wanted. You'll find a bad reputation is the hardest thing in the world to change.

NO ONE VALUES THE SHAME WORKERS

Think of it from the point of view of the person asking you to do the work. They're probably not totally unaware they're asking you to do some shady shit. They've probably convinced themselves it's a stop-gap. A temporary bit of shade to get the company back on track, perhaps. For example, I doubt Uber designed Greyball[1] because they wanted to be evil. My guess is they imagined it was

---

1 Greyball was an internal tool used to identify and circumvent officials who were trying to clamp down on the ride-hailing service's predatory growth techniques. Uber flagged phone numbers tied to city officials and regulators, then served them a bullshit map showing there were no cars available.

a necessarily small evil that helped them achieve a greater good. (You can justify anything if you try hard enough. Or just want to.) But in the end, everyone associated with that project is covered in shame. You were a means to an end they never want to see or think of again.

This is not the path to respect. And it is not the path to a long career. This is a path to a career doing short stints of shame work.

WHAT THE FUCK MAN I JUST WANNA DESIGN STUFF

I get it. You like to make things. You became a designer because you enjoyed designing. I did too. But there's more to this job than that. We work within a tight, fragile ecosystem where our labor has repercussions. And you are lucky enough to be a designer at a time when design is taken seriously and when design has power. With that power comes responsibility. You are responsible for what you put into the world. And you are responsible for the effect your work has on the world. And right now designers—I define this term broadly, by the way. Perk up your ears developers and engineers—are creating new inroads in all manner of things. We're designing software for self-driving cars. We're designing software which intimately touches people's lives. We're designing software which puts people in strangers' cars. We're designing databases that track immigrants for eventual deportation.[2] Some of these things need to be designed with the strictest ethics in mind. Some of these things don't pass an ethical test and shouldn't be designed at all!

So, I get that you like making things. But making things at the expense of someone else's freedom is fucked. Not putting what you're designing through an ethical test is not just lazy, it's dangerous. Feigning ignorance, claiming that ethics is not part of your job as a designer, is no longer a valid stance. Knowing that ethics are part of your job and ignoring it is criminal.

THE REAL QUESTION

A better question than how you're going to pay your rent working ethically might be why you are even open to behaving unethically? Look around at the other professionals you interact with on a daily basis. Your doctor. Your grocer. Your mechanic. Your congressperson! How would you react to knowing they're entertaining doing their job unethically? Think of the ones you'd be

---

2 Those same databases were eventually used to round up immigrants and separate them from their children. As of this writing, there are 666 children in US custody whose parents can't be located. If that number seems apt it's because the Trump administration is the fucking devil.

appalled by. And the ones you expect it from. And your relationship to the people on both those lists. I don't want designers on the same list you just put your congressperson on.³ I'd be honored to be on the same list as your butcher.

---

3 Bernie, AOC and the rest of The Squad excluded, of course.

IT MIGHT BE TIME TO START FLIPPING TABLES

(Originally published in Modus on July 12, 2019)

Q: In your last column you talked about how to persuade your clients and employers to do the right thing by building a good argument and alliances. But what happens when that's not enough? What happens when they insist on doing unethical shit?

Well, we're eventually going to get to the point where we flip some tables over. I just didn't want you thinking that was your first course of action. But there comes a point, after you've exhausted all other courses, where you have to put your foot down. There are things you cannot do. Case in point:

On July 3, 2019, U.S. Customs and Border Patrol of Arizona posted a video tour of one of the concentration camps along our southern border. I've watched it, and if you want to watch you can Google it; just prepare yourself—it's very triggering. It's a video that justifies jailing refugees and separating children from their parents. It's a video that lies about the conditions these human beings are living in. Its goal is to make you feel okay about the racist and dehumanizing acts our government is perpetrating. It's totalitarian propaganda.

But hey, this is supposed to be a column about design advice. So why are we talking about this? Well, one, because we should give a shit. The fact that we have concentration camps in our country is something that should piss us off, and we should take action against it. I have no doubt many of you are with me on this.

So let's move on to issue number two.

Shortly after this video was released, RAICES Texas, a nonprofit doing amazing work providing educational resources and legal aid for migrants, discovered the video had been created by Ogilvy, one of the biggest and most renowned advertising agen-

The Righteous Angers

cies in the world. Raices tweeted out "@Ogilvy has a $12m contract with CBP to help them with public relations such as this video."

A few days later, Ogilvy denied they'd had anything to do with the video. And CBP claimed the video was produced in-house. I'm going to have to take both of them at their word, although I talked to a few people who've worked in government agencies and they were very doubtful CBP has an in-house team capable of producing a video like this.

For Ogilvy's part, although they denied producing this particular video, they didn't deny they've contracted with CBP for PR work since at least 2017. They're the agency of record. And since multiple contracts have been signed since September of 2018, after we were all aware of what was going on at those camps, if not the extent of it, Ogilvy certainly can't claim they weren't aware of the kind of work they might be associated with as the agency of record on the account. Being blamed for a PR video a company puts out while you're that company's agency of record is a fair assumption.

So Ogilvy was either unaware of the video (at which point you'd wonder why a client went elsewhere to do the work when they had an agency on the books), or they passed on the video (again, they must have known what they were getting into when they signed that contract), or everyone's lying. An accusation of which, obviously, I have no proof. And it would be weird to accuse a PR agency of lying. So, I won't.

The thing I'd rather focus on isn't this particular video. It's the contract. Ogilvy knowingly took a contract to work with CBP after ground had broken on the camps. That's the bigger story.

Here's a little history: Ogilvy was founded in London in 1949 by David Ogilvy, the "father of advertising," a title he probably gave himself. He wrote two very influential books about advertising: Ogilvy On Advertising and Confessions of an Advertising Man. The company grew worldwide, changed names a bunch, sold a lot of Dove soap, Rolls-Royces, and oil. David Ogilvy stepped down as chairman in 1973. Since his departure the company has had its fair share of success under various leaders, among them Paul Hicks, who rose to regional CEO of the Americas. You may not be familiar with Paul Hicks, but my guess is you've heard of his daughter—Hope Hicks, former White House communications director under the Trump regime. As Lestor Freamon would say, follow the money.

And through this combination of prestige and influence, a group of people—designers, writers, directors, photographers, and project managers among them—who all happened to work at Ogilvy, ended up being told their next client was the U.S. Customs and Border Patrol. Who were building concentration camps. And that their job was doing PR for them.

Did any of those people try to talk Ogilvy out of the taking on the client? At this point we don't know. And we may never find out. But perhaps they did. Perhaps more than one of them did. Perhaps there's a whole bunch of people at Ogilvy pissed off about being associated with this client. Perhaps someone even stormed out and didn't want anything to do with this client. We may hear from them yet. (Unless of course, Ogilvy bought their silence with a big fat settlement.)

Here's what we do know: A propaganda video was made to cover up atrocities happening in concentration camps along our southern border. That video couldn't have been made without several people's labor. Paychecks were cashed. And if you helped make it, that particular video may go in your book, or portfolio, or reel, or whatever you call it, or it may not. But its inclusion doesn't justify what you did, and its exclusion doesn't forgive it.

Enough people were willing to use their talent and their labor to get that video made, and I assume they were paid handsomely for it.

THE EMPLOYEES' RIGHT TO VETO

Where you put your labor is a choice. Always. Sometimes it may be a difficult choice. For example, if your employer controls your healthcare, and you or someone else covered by that healthcare have a medical issue that needs coverage, you're going to weigh that when you make the choice of whether to work on a project or not. But you're still making a choice. (This is one of the many reasons we need to decouple health coverage from employment.) I can't begrudge someone choosing their family's health. But I can begrudge them saying they had no choice. They did. And they made the one they needed to. I would, however, counsel you to move to a job where you're not forced to choose your family's health and well-being over other families' health and well-being.

One of the cornerstones of our design studio, Mule, was that we never wanted to take on work we couldn't be proud of. If it was worth doing, it was worth telling everyone we did it. No secret clients. That meant two things: We needed to stay small enough that we could be picky, and we gave every employee veto power on the jobs we took on. Everyone working for us knew they could veto a project from their first day of work. And staying small meant that every person was critical to what we did. If someone didn't want to work on a project, we couldn't afford to shield them from it. We just didn't take it. We also hired people who had opinions and weren't shy about them. And honestly, I think that's one of the reasons people worked for us. They knew they'd never have to work on something they were ashamed of.

So, before you take your next job, ask if employees have a right to veto work. Ask whether you have a say in the type of work you take on. And talk to members of the current team to make sure this is the case. It's not too late to do this at your current job, either. Negotiate as a team. Let your employer know that you want to discuss every project that comes in before it's signed. Is every employer going to be open to this? No. But that's a good thing to know as well.

THE COST TO YOUR CAREER

Very few of us are in the last job we'll ever have. At some point you'll move on. You'll want to do something new, or the place you're at now might close up. You'll be on the hunt for a new job. You'll polish up that resume and find yourself across the table from a prospective new employer. They'll go over that resume. Maybe they'll care that you were at Ogilvy in 2019. Maybe they'll ask about the propaganda video you made. Maybe they won't. Maybe they'll care you were on the Twitter Trust and Safety team at the height of abuse and harassment. Maybe they'll ask why you didn't do more to stop it. Maybe they won't. Maybe they'll care that you were on the Superhuman team that tracked every time someone opened an email and then geo-tagged it. Maybe they'll ask why you were okay implementing that. Maybe they won't. Maybe they'll care that you were on the Uber team that developed Greyball, a tool that deceived regulators. Maybe they'll ask why you thought it was ethical to do that. Maybe they won't.

But maybe they will.

Ultimately, the reason we should work ethically shouldn't come down to how it affects our career. But since much of the resistance to working ethically is, sadly, about career impact, it's worth considering the longer-term effects. Doing unethical work might keep you safe in your current job; it might even help you climb the ladder. But fuck, what is it doing to your career?

With each unethical decision you make, the path you're charting through your career changes. Paths close. And the paths left open don't always look too good. Covering your ass in the present is only exposing it to the future.

And reputation matters. Someone who succeeds by doing the shitty work that more ethical people won't do is going to be known for that. A lifetime of following orders rarely leads to a career in leadership.

ALLEGIANCE TO A COMPANY IS STUPID

My dad was a construction worker most of his life. And we grew up in a place where the ground froze over every winter, which meant he got laid off every winter. Sometimes, once spring came back around and the ground thawed, the same company that laid him off would hire him right back. And sometimes they wouldn't. He'd move on to another company. The companies he worked for had no allegiance to him. They'd dump him off payroll as soon as they needed to. To be honest, he didn't have too much allegiance to them, either. He sold them his labor. And he was proud of the work he did, but he took pride in the work, not in the company that did it. The companies were trying to build as fast as possible and profit as much as possible. That is what they do. The workers were building places where people took shelter. Where they lived. Where they worked. Where they went to school. And they wanted those people to be safe. So they did their jobs well, for the people using those buildings. And for their own reputation. Because your reputation got you your next job.

And when the ground froze, my dad would put away his tools for the winter. We'd decide which of the rooms in the house we could afford to heat. The union would bring groceries around. We'd stretch out the unemployment checks as much as we could. And small loans would fly back and forth among all the laid-off construction workers. Regardless of where they worked, those construction workers always had each others' backs. Their allegiance was to each other. Not to the companies they worked for. Because those came and went.

When the ground thawed, the foremen would get called back first. Then they'd build their teams. Your reputation was important. If you had a reputation of doing good work, you got called back. If you were a shitty worker, if you put the company's needs above the work...well, you weren't called back. That road was closed.

If you're going to have an allegiance to something, make it the people who need to live with your work. If you're going to have an allegiance to two things, make it the people you work with.

The problem with loving where you work, is that you start believing they love you back. They don't. Everyone gets laid off when the ground freezes.

Oh, and for the record, I called my dad and asked him if he would've taken the job building the camps. "Fuck no. All our guys were immigrants."[1]

---

[1] A year after telling me this, my Dad cast his vote for Donald Trump, proving yet again that you should never attempt to make your parents the heroes of your story— and that immigrants are just as capable of being racist as anyone. Sometimes more.

IGNORANCE BEATS EMPATHY

(Originally published in Dear Design Student on December 8, 2015)

> Q: I'm trying to be more empathetic toward the users of the things I design. Do you have any useful advice?

Empathy is a wonderful tool.[1] Not just in our design work, but in life. It allows us to understand things from other people's point of view. It means we make an attempt to walk in their shoes. See things from their perspective. To put it bluntly, empathy is one of the things that keeps us from being total assholes. It's also how we double-check that our work isn't falling into some ethical or moral black hole, by attempting to see how it affects people who aren't us.

As designers, empathy means we consider the people we design things for. We consider how their interaction with what we make will affect their lives. We consider how well they understand what they are using. And we attempt to be honest about what kind of experience they'll have. There are several tools designers have developed over the years to help us gain a sense of empathy toward our users, most notably personas, where we attempt to create "real" users. It's kind of like method acting.

But at the end of the day, we are who we are. And our empathy only stretches as far as our experience can take it. We don't know what we don't know. And even if we feel bad for that person who lives on the street, who probably needs healthcare, dinner, and a place to sleep, we only feel bad about it for a few steps on our way home. And then we return to our lives.

And as empathetic as we might be, we are hopefully aware

---

1 *Most corporate empathy workshops are bullshit. If you want to know how women, or BIPOC, or neurodiverse people think, you don't hold a workshop so that your all-male, all-white team can learn to think like them—you hire them.*

The Righteous Angers

that the world worked out pretty well for us. After all, empathy is expensive and only available to you if you have the time to feel it. Other people are too busy surviving. And we have to admit that the world, as designed, works more or less in our favor. (Yes, I am fully aware that I am a white dude writing this. And that our mileage varies wildly.) So as much as we talk about disruption, we don't really want to disrupt things too much.

But what if that wasn't the case? What if you woke up one day and found yourself on the other end of the stick? Let's get philosophical about it...

You're less likely to advocate for slavery if there's an actual chance that you might be one of the enslaved.

In 1983, the great American philosophers Randolph and Mortimer Duke ran a study where they switched the lives of two men, one well-to-do, the other down-on-his luck, to see whether a world designed to succeed for one would also succeed for the other. They intended it as a study of nature vs nurture, but interestingly enough once the men's positions were reversed back, the once-again well-to-do man, now having been at the short end of the stick, started making different decisions. Shaken in his entitled beliefs, he was now designing a world where he was unsure of what position he might be in tomorrow. Obviously, we are talking about *Trading Places*. One of my favorite movies. And a wonderful introduction to both Eddie Murphy, and a philosophical concept called the veil of ignorance.

What's a veil of ignorance? Great question. In short, a veil of ignorance is a way of determining whether something you're making sucks by allowing that you may end up in any possible relationship to it. The concept was developed by an actual great American philosopher, John Rawls.

Here's Wikipedia's excellent example:

> *...for a proposed society in which 50% of the population is kept in slavery, it follows that on entering the new society there is a 50% likelihood that the participant would be a slave. The idea is that parties subject to the veil of ignorance will make choices based upon moral considerations, since they will not be able to make choices based on self- or class-interest.*

Get it? You're less likely to advocate for slavery if there's an actual chance that you might be one of the enslaved. Now, I doubt anyone here is an advocate for slavery. So let's bring it back to things more in our wheelhouse.

Imagine, if you will, that the leadership team at Uber woke up tomorrow to find out that they were no longer the leadership team at Uber. Instead they were now drivers. And not only were they now beholden to the decisions they'd made as the leadership team, but they were also unable to change those decisions. Because those changes are above a driver's pay grade.

Or imagine that the leadership team at Airbnb woke up tomorrow as a lower-class immigrant family being served an eviction notice because their landlord did the math and decided he could make more revenue by renting their apartment out as a short-term rental. Would they wish they could revisit some of the decisions they'd made as the leadership team of Airbnb? I hope so.

Or imagine the teams that manage online harassment at any number of social networks. Imagine they wake up one day, no longer working at those companies, and being contacted by former stalkers or abusers. Imagine them having to go through the same process as so many of their users do every day.

If any of these groups were given the chance to go back to their old lives for just one day, I guarantee you they would design their products differently. And that's what a veil of ignorance is all about. It's about designing a system where you could ultimately be getting the shit end of the stick. Empathy is about trying to put yourself in other people's shoes in an existing system. The veil of ignorance helps you create a just system, which is a different thing. And it might cause you to design things just a little bit differently. Just a little bit more fairly. It's the single most important political and ethical concept in a designer's toolbox.

It's not enough to disrupt a thing, you have to disrupt it the right way. You have to design the right thing.[2]

---

[2] *By the way, this short little essay became the take-off point for a lot of the things I'd write about for the next five years, including my book Ruined by Design. You can see a lot of the bits in there kinda start to bubble up awkwardly in this piece. By itself it's kind of meh, but I'm leaving it in to show you that sometimes you start to scratch the surface a little bit and you gotta put things out there and wait for them to come back more fully formed—at which point they'll consume you.*

POLITICS IS THE DESIGN PROBLEM OF OUR TIMES

(Originally published on Medium on March 1, 2019)

A couple of years ago I was talking to a conference organizer about doing a talk. Doesn't matter who it was.[1] That's not the point. Everything was going well until they asked me the name of the talk.

"How to Fight Fascism," I replied.

"Could you do something less political?" they replied.

I ended up not doing that conference, and that story repeated itself a few more times since then. People want to hear about design, and they agree that we might have designed ourselves into a "bit of trouble." We may even be coming around on the idea there might be some people in Silicon Valley who don't have our best interests in mind. But for the love of god, can we please address it without getting political? (After all, it upsets the sponsors.)

If we're going to talk about design responsibility, ethics, or whatever you want to call it, we're going to have to get political. Let me show you why. Let's walk through a few scenarios—the names are fake, the situations are not:

Tom works at a very large social media company. He's good at what he does. That's how he got the job. He's not in charge of any projects or managing anyone, but he's a valuable member of his team, which is designing the system for buying ads. That's an important system for the company, it's how they make their money. It's gotta work. It's also complex. There's a lot of knobs and levers to manage. All of those levers control who the buyer can target with the ad. Targeted advertising is important to the company, the more targeted the ad, the better it converts. Tom's job is to make sure all the levers are working, as well as to add and subtract levers as management requests. He gets a request to add a lever for religion. He

---

1 *It was Adobe.*

reads the spec. It says that buyers should be able to select from a list of religions so that only people who self-identified as being part of that religion receive the ad. Because Tom is good at what he does he runs through the possible scenarios this new lever introduces. One of the scenarios Tom comes up with is the possibility of that someone might check all the religions except Jewish to send anti-Semitic ads. He raises this as a red flag to management. Management waves it off. Tom considers pushing the issue, but then remembers that his entire family counts on him for the healthcare he gets through his employer. Including his daughter, who had a medical scare last year. Tom can't put their care at risk. The feature gets built.

Agnes works at a ride sharing company. She's good at what she does. That's how she got the job. She's also the only woman on her team, which is currently working on a tool to track riders. The product spec says the company wants to keep tracking the rider's location at all times, regardless of whether they'd just finished a ride or not. Agnes doesn't agree with this, so she takes it to management. They wave her off. Agnes considers pushing the idea, but then she remembers she's paying off $100,000 in student debt. $350 a month. The amount seems insurmountable. Agnes relents and helps build the feature. Maybe public outrage will get the feature deleted.

Nigel also works at a social media company. He's on the Trust & Safety team, which is responsible for making sure the platform is free of abuse and harassment. He spends his day reviewing content that users have marked as harmful. He notices a large number of complaints about the same user and once he investigates, he realizes that user has a history of harassing other users. He takes this to management. Management doesn't deem the harassment bad enough to suspend the user. They overrule Nigel. Nigel disagrees and considers pushing the issue but then remembers that he's in the country on an H1-B Visa, sponsored by the company. If he loses his job he'll be deported. Nigel doesn't suspend the user.

To be clear, doing the right thing means doing the right thing. No matter the cost. But fuck, can we make it a little easier on people? I can't fault someone for choosing their family's health over an ethical dilemma, but we can work to take that dilemma away. No one should ever have to choose between their family's well-being and working ethically.

So, tell me again... how are we going to talk about design ethics without getting political? We're dealing with healthcare. We're dealing with student loan debt. We're dealing with immigration reform. That's shit we need to fix at the polls.

It's political. Thinking otherwise isn't just foolish, it's irresponsible.

HOW TO FIGHT FASCISM[1]

(Originally published in Dear Design Student on November 30, 2016)

Q: I am freaking the fuck out that America just elected a fascist. I'm bouncing between wanting to fight and hiding under my bed. I'm just a designer, what can I do?

Hi. Take a deep breath. I've been breathing into a paper bag for three weeks myself. Let's not whitewash (ha!) this. We are truly fucked. We're standing at the very edge of the American experiment. I can't blame you for not wanting to take that next step.

So, let's take stock: barring any last-minute Hail Marys such as vote audits, appealing to faithless electors, or praying that Congress actually gives a shit about conflicts of interest, Donald Trump is set to become the President of the United States. We should behave as if he's going to do the things he's said he's going to do. There's no secret liberal inside the orange jumpsuit. And while he may not be a fascist himself, he's a needy narcissistic Zelig-like sociopath who needs to be loved and admired by those closest to him. And he's chosen to surround himself with fascists.

In the next four years, civil rights in America will be under constant massive attack. As designers, we don't get to opt out.

---

1 As you can see from the original publication date, this essay was written shortly after the 2016 presidential election. After publication, most of the feedback I got was that I was being "histrionic" and overreacting. Coming back to it four years later, it feels reserved. Not even at my most "histrionic" did I picture babies in cages and concentration camps along our southern border. I'm writing this footnote shortly after the 2020 presidential election, just as Trump is refusing to concede his loss and fabricating stories about massive voter fraud. I eventually turned this essay into a talk. You can Google it. Sadly, I ended up doing the talk mostly overseas, as design and tech conferences in the US didn't want anything to do with it, including Adobe, as I mentioned in the last essay. I may have an axe to grind. I'm not petty, I'm from Philadelphia.

And since you've all been screaming about changing the world, now's the time you realize it's not done by disrupting viral video consumption delivery systems, but by actually getting involved in some of that civil shit.

GET HEALTHY

First off—and I cannot stress this enough—get your shit together. Your mental health is important. If you're not taking care of yourself, you're in no position to take care of others. And right now, not taking care of yourself is the most selfish act possible because we need you. So, go to your therapist, get back on your meds, do some physical activity to get those endorphins going because we need you strong and we need you healthy. And look, if you're seriously freaking out, know that I totally understand why you would be, because this all sucks. Then know I love you. Then call this number: 1-800-273-8255. They will take care of you. If you need some help figuring all this out, but can wait a few days, you can find someone to talk to here. Please take this seriously. Once you're healthy, we can get to work.

BE A CITIZEN FIRST

Before we need you as a designer, we will need you as a citizen[2]. As a citizen there are three things you can give: time, money, and haven. Give what you can of each.

The most important thing you can do is make phone calls. Call your representatives. Call your senators. Do not tell them about your feelings. They don't care about your feelings. They care about your next vote. Tell them what they need to do to get it. Be relentless. Be the person who calls every day. You can make 5 calls a day and you'll be in and out in under 15 minutes. And yes, I mean phone calls. That's how these people work. A call is ten times more valuable than an email, or a petition. You all have phones. You're probably reading this on one.

LOOK FOR THE HELPERS

There are organizations out there busting their ass to keep the flame of democracy alive. Find them. They need your money to keep doing this. Give to the ACLU, the Southern Poverty Law Center, Planned Parenthood, the Anti-Defamation League, the Trevor Project. There are so many more.

---

2 *I regret the use of the word "citizen" here because it excludes so many people who did the work despite not being citizens, including people who had their path to citizenship taken away. Perhaps a better way to phrase it would've been "be civic-minded."*

Help those still willing to speak truth to power. The media failed us in a big way, but there are enough think pieces about that already, and this is not that. Donate generously to the ones still standing proudly: Mother Jones—still fierce after so many years and stronger than ever, ProPublica—going deep into the data to expose corruption, The Washington Post—still standing tall. And, oh how I love adding Teen Vogue to this list. Because they're kicking a lot of ass, and they're talking directly to the next generation. The New York Times has already fallen.[3]

Mostly, though, as a citizen, look out for each other. And especially look out for people most likely to be the target of roving jackbooted deplorable thugs. There is more safety in numbers. We're past the point of pretending not to see that hate crime happening on the bus. It's happening. And it will keep happening until the people on the bus, or on a plane, or in the street, stand up to the haters. Is this putting your safety at risk? Yes. Is it increasing the overall safety? Yes. If we're going down, we're going down as a community.

THERE'S NO WRONG WAY TO HELP

Some of us are fighters. Some of us are check writers. Some of us like to protest. Some of us are focused on vote audits. Some of us are preparing for four years of absolute terror. This is all good. And this is all necessary. Don't tell someone they're helping wrong just because they're doing something you don't agree with. (Unless you think it actually runs contrary to their goals. Then it's called critique, and is allowed. Sometimes.)

If it gives people hope it's worth doing. Hope is gonna be in some short supply in the next four years so let people generate it where they can. If you're organizing a protest, but your friend Bob would rather sit at his desk and make phone calls, that's cool. He's doing his thing and you're doing yours. The amount of time we spend arguing with each other only subtracts from the amount of time we spend fighting fascists.[4]

GO LOCAL

Fix yourself. Fix your house. Fix your street. Fix your neigh-

---

3 Amazingly, they fell even further in the next four years.
4 One of the lessons I hope we learned from dealing with this garbage administration is that there wasn't one thing that took them down. It's like we went years looking for that perfect smoking gun that was going to bring Trump down, from emoluments, to the Mueller case, to impeachment. In the end, it took everyone doing everything at once. It took the check writing, the door knocking, Stacey Abrams mobilizing people in Georgia. It took the TikTok kids. It took a pandemic. It took everything at once.

borhood. Fix your city. Fix your state. Fix your country. That's the order to work in. Start small and work up. This election was lost at the local level. We won it at the national level!

Get a grip on what's going on in your local community, and find out where you can help. Maybe it's keeping your street free of swastika graffiti. Maybe it's making sure the two Muslim kids down the street have a clear path to school and back every day. If you own a business, make sure everyone knows they're welcome there. Put up signs. Make them visible. Make sure that any fascist walking into your place of business feels unwelcome.

Who are your local elected officials? Where do they stand on things? When are their terms up? Local officials turn into state officials. State officials turn into national officials. (Unless of course, we keep electing idiots who run golf courses. Badly.) Weed the bad apples out locally so we never have to deal with them nationally.

And I can't emphasize this enough. The shit on the ballot that actually affects you is that long winded badly worded local shit. (It's designed that way on purpose, by the way.) That shit has an almost immediate effect on your community. And as a local voter in a smaller voting base your vote has a higher percentage of mattering. So read it; try to understand it. Someday you'll be so far up in it, you'll be able to influence it.

DON'T WORK WITH FASCISTS

We do not work with fascists. There is no reason to reach out to fascists. We don't build bridges to fascism. We burn down the bridges which link them to us. These are people who think those who don't look like them as subhumans. I have no desire to reach across the aisle to that deplorable vomitous shit. If you help them you are advancing their agenda, and become no better than they are.

WORK ETHICALLY, NOW MORE THAN EVER

More than ever we need to look at how we're designing the world. We also need to look at who is designing it. Don't look for much resistance from Silicon Valley, which is run by rich white boys, for the benefit of a government run by the rich white men they aspire to become. Take stock in who you're working for, what they're making, and who they're hiring to make it.

We have serious problems to solve. Make sure you're working at places that are interested in solving them. And if you're working at large companies, especially companies in the social space, keep an eye out for how your products affect the marginalized. Work on ways to empower those who need empowering. And if you're in the "news" business, maybe take a look at what kind of lies you're spreading. (Yes, I'm talking about you Facebook.) Take a look at the words you're using. Stop sugar-coating shit. Don't say alt-right when you're talking about Nazis, dammit. (Yes, I'm talking about you, New York Times.)

As designers, we need to fulfill our mission of being gatekeepers. Your job is to improve the world for everyone, not just those in power. When you are asked to work on something which might marginalize people, you must stop. When you are asked to come up with solutions which tighten the grip of fascism on your community, you must stop. And not just move over so that someone else can take the oar, it is your responsibility as a designer to make sure that work never sees that light of day. Even if that means throwing your body across the gears of fascism. This is the job. This is what you signed up for.

And the company that signs on to build the Muslim database can go fuck themselves.[5]

PREPARE FOR A BETTER FUTURE

We're in this situation because we designed the world to work this way. The rescinding of the Voting Rights Act, the gerrymandering of congressional districts, the existence of the electoral college, the underfunding of schools to create an undereducated electorate. We did all that. And we'll have to undo it. This will take a fuckton of time, which honestly I'm not sure we have. But we have to try.

Right now, as we speak my friend Dana Chisnell is working on how to improve our election process at the Center for Civic Design. Part of me thinks she is insane because we're never going to have another election. But I'm trying to listen to the part of me that wants to ensure she has a chance to get there.

So we fight. We fight because we can't not fight. We fight because maybe this is the cliché darkest before the dawn. We fight because if we don't people, get beat up, rounded up, stripped of their dignity, and killed. And while this may or may not happen in a large government-sanctioned way, it's definitely going to happen in small pockets of deplorable misery throughout the country. And there are things we can do to prevent it. We have to try.

Fight fascism.

---

5 *It was Palantir.*

IN PRAISE OF THE AK-47

(Originally published in Dear Design Student on July 27, 2015)

> Q: I had a professor in school who went on and on about how well the AK-47 was designed. He stressed that as designers we should be able to appreciate an object's design on a purely aesthetic level. Do you agree?

Fuck no. Fuck him. Fuck the AK-47. Fuck all guns, and the people who design them, but especially fuck Mikhail Kalashnikov, the designer of the AK-47.

Let's look at the argument being made. The AK-47 is often cited as a well-designed object. And this case is usually made by pointing out that the AK-47 is easy to use, maintain, take-apart, modify, and manufacture. It's a model of simplicity. And the original design, introduced in 1948, is still in use, even as the AK family has continued evolving.

> ...the model and its variants remain the most popular and widely used assault rifles in the world because of their substantial reliability under harsh conditions, low production costs compared to contemporary Western weapons, availability in virtually every geographic region and ease of use.
> —Wikipedia

Any of us would be proud to design something with that kind of legacy. Ease of use. Ease of manufacturing. Adaptability. Simplicity. Aren't these the cornerstones of design?

So, where is the problem? Surely, a designer's job is to design something to the best of their ability. As a designer, you are required to do your best work. And we've all had to design something we weren't too crazy about. In which case your responsibility is to improve the design. How do you improve an object that's designed to kill without making it more efficient at killing?

A gun's only purpose is to kill. When it kills, it is working as designed. And a gun is designed to be fired. The trigger yearns to be pulled. It is designed to shoot a bullet into a human body at a force that creates the maximum amount of damage. Which is technical way of saying its job is to kill you.

But while someone can certainly make the case that an AK-47, or any other kind of gun or rifle, is designed, nothing whose primary purpose is to take away life can be said to be designed well. And that attempting to separate an object from its function in order to appreciate it for purely aesthetic reasons, or to be impressed by its minimal elegance, is a coward's way of justifying the death they've designed into the world, and the money with which they're lining their pockets.

And yes, there are many objects that kill. Cars come to mind. And they're the gun enthusiasts favorite straw man. And while I agree that cars definitely have the potential to kill, you can't really argue that they're designed to do so. Car deaths—and I hesitate to call them "accidents" because I do believe there are too many of them—are a very unfortunate by-product of car usage, but not the main goal. Every year steps are taken to make cars safer, to improve the design of cars to reduce the amount of deaths. (Along with other, more marketable, goals.) But, by definition, improving the design of guns can only result in them becoming better killing machines.

If a thing is designed to kill you, it is, by definition, bad design.

What is the designer's role in this? Design is an ethical trade. And yes, it is a trade done for money. But we have a choice in how we make that money. A designer possesses a set of skills necessary to get something made. And needs to properly assess how they are putting those skills to use.

But, won't someone else just design it?

Possibly. If Kalashnikov hadn't designed the AK-47, wouldn't someone else just have designed another rifle? Most assuredly. And they did. There are as many types of rifles out there shooting up our villages, our churches, and our Marine recruiting stations, as there are cereal boxes in the cereal aisle. And they all have a designer's name attached to them. The shit we design carries our name.

Your role as a designer is to leave the world in a better state than you found it. You have a responsibility to design work that helps move humanity forward and helps us, as a species, to not only enjoy our time on Earth, but to evolve.

And to design is to take purpose into account—as my friend Jared Spool says: "design is the rendering of intent." You can't separate an object's function from its intent. You cannot critique it, you cannot understand it, and you cannot appreciate something without thinking about its intent.

You are responsible for what you put into the world. And how it affects the world.

You are responsible for what you put into the world. And you are responsible for how what you've designed affects the world. Mikhail Kalashnikov is responsible for as many deaths as the people who pulled those triggers.

Obviously, firearms design is an extreme example of this. I doubt many of you will go on to become firearms designers, and fuck all of you that do. But how many of us are asked to design things which have the potential of causing harm to the people who come into contact with our work? How many of us will work on privacy settings for large social networks at some point? Will we think of how those settings affect those who interact with them? How many of us will design user interfaces for drop cams? Will we think of the privacy violations they might cause? How many of us will design products that put people in strangers' cars? Will we consider those passengers' safety as we design our solution? And will we see it as our responsibility to make sure these products are as safe as possible?

And if we come to the conclusion that these products cannot be made safe, how many of us will see it as our responsibility to raise our hands and say "I'm not making this."

Because we have to.

THE HAPPIEST ARTICLE YOU WILL EVER READ
ABOUT DESIGN ETHICS

(Originally published in Modus on July 25, 2019)

> Q: Dude. I get that it's important to work responsibly and ethically, but every article I read about ethical design makes it sound like punishment. What a fucking slog. You make it sound like working ethically is all hair shirts, self-flagellation, and crying. Why are all the articles about ethics so negative? Can we please get something positive for a change?

You know what? You're not wrong. I'm certainly guilty of this. And I tend to come into conversations pretty hot. Obviously, design ethics is a serious issue. But I do tend to dip my quill in more vinegar than sugar when I write about it. Why is that? Maybe because I want to convey the seriousness of the situation we're in? Maybe because I enjoy writing angry? Maybe I'm just a grumpy old man—who knows? Tell you what? I accept your challenge. Today I am going to write the happiest article you will ever read about design ethics.

I LOVE DESIGN

I'm going to start by telling you how much I love design. Because when you get down to it, that's where all this is coming from. That's hopefully why we're all here. We're all united together in this thing we call design. And yet, we have at least as many definitions for design as we have for god and for what makes a good pizza. For some of us, design is a thing we can see, for others it's what we can't see, and for yet others it's the plan that puts everything in motion. For some of us good design is an intuitive interface, for others it's a well-designed football play, or the Dead Kennedys logo, or a Robert Altman plot, or an activist action. And somehow none of us are wrong. But so much of our ability to work together involves finding people with definitions of design that match—or, even better—complement each other, and a willingness to learn from those we don't understand.

And yet, I believe there's a definition that any of us can stand behind: The whole spark of design comes from our seeing another human being attempting to do something and not having a great time doing it, and thinking, "I bet there's a way to help them do that." And then figuring out what that is. That's it, that's the whole thing. Whether we're helping someone fill out a naturalization form in their own language, keep better track of their finances, find the right charity to donate to, keep in touch with someone they love, or just find the right cat gif, everything we design should be in service of making people's lives easier. Feel free to replace "easier" with "more efficient," "delightful," or any other positive word you prefer.

And sure, there's a million ways to solve that equation. There's a ton of methodologies around it. But for all we might argue about "whose methodology is better" online, it's worth taking a minute to realize that if we boiled all those wack-ass methodologies down, we'd eventually get to the same core equation: Design is helping people figure out a better way to do stuff.

You gotta love two things in this business: design and people. And you should love people more than you love design.

So, I love design. I love helping people. As much of a grumpy old man as I am (and I am) I love that I get to spend my life helping people do shit. And now that I've been doing this design thing for a while, I even get to help the people who want to help people learn how to do it better. I like doing that even more.

And of course it follows that I cannot love design that tricks people, or traps people, or separates people, or keeps people confined.

Design! Kumbaya.

I LOVE DESIGNERS, BUT NOT AS MUCH AS I LOVE DESIGN

Design is done by people. All sorts of people. Anyone who's figuring out how to solve a problem, within a set of constraints, to make someone's life easier (their own included) is designing something. So is everyone a designer? Let me put it this way: Remember Chef Gusteau's book *Anyone Can Cook* in the movie with all the cartoon rats? It's true—anyone can cook, and some people are good enough at it to get paid for their effort. So yeah, anyone can design. And a few of us are good enough to get paid for the effort.

A chef who gets paid for their effort has a responsibility to feed people the best food they can cook, and they also have a responsibility to feed them food that won't make them sick. So it is too if you're a designer who's paid for your effort. You have a responsibility to design things that make people's lives easier. And if the people you design for are on the other side of a screen, they'll never see you—much like most chefs never see the people they're cooking for—but you find joy in a job done right. And in knowing that those people are better off for having experienced your work.

I love designers, but not as much as I love design. I love that you want to help people. And in keeping with our equation, I want to help you do what you do better. As an old man in this business, I want to clear a path for the younger folks coming up behind me. Not because I'm smarter than they are. On the contrary: many have shown that they're smarter than me. I want to help them because I've got over twenty years of lessons to share. Over twenty years of seeing patterns behind how people behave (including clients and bosses), over twenty years of seeing the same problems emerge again and again, over twenty years of making decisions both good and bad, and over twenty years of figuring out how to avoid obstacles. And because young designers are smarter than I am, they can decide which of these lessons are actually valuable to them or not.

If you're lucky—and I have been— the up-and-comers end up teaching you a lot more than you can teach them. They come into the field with new perspectives, with more diverse backgrounds, with experiences different from my own. It's breathtaking. And humbling. In the best way.

I LOVE PEOPLE WHO AREN'T DESIGNERS MORE THAN I LOVE DESIGNERS

Back to our chef analogy: Before you start throwing ingredients in the pot, I want you to walk out into the restaurant and look at those hopeful hungry people out there. Some of them are on their first date, some of them got a sitter and are having their first quiet meal together in a year, some of them are having a memorial dinner in honor of a friend who passed, and some of them just had a really shitty day, which this meal could really help them ease out of. Those are the people you're cooking for. Those are the people to whom you owe your allegiance. So when the restaurant owner tells you to skimp on the portions, or to use yesterday's fish, or to water down the liquor a little bit for the sake of profit, you should tell the restaurant owner to kiss your ass. Because you are here to give people the meal of their lives.

Is the restaurant owner going to be upset about that? Possibly.

Might he fire you after you pull that stunt a few times? Possibly. But here's the thing: Go back out and look at those people waiting to eat. You're cooking for them. They're eating your food. You want them to have the best possible meal. That person who had a shitty day? Your meal is gonna make the difference for them. That couple on their first date? Your meal might just be the reason they decide to have a second date. The table having a memorial for their friend? Her memory deserves a great meal. The parents having a night out? Well, they may just name the child they conceive tonight after you, if they aren't too full after that amazing dessert. Maybe tomorrow.

Right now, the biggest problem in design is figuring out how to give designers a better way to do the stuff they need to do. This means reminding them of how much collective power they have, how much responsibility they bear to look out for those interacting with their work, and building an infrastructure which allows them to do this work without fear of repercussions.

A chef who takes care of their diners gets a reputation for taking care of their diners. They move on to bigger jobs. They get more staff. Their career is on a trajectory, and that trajectory is greased by happy people with full bellies and good memories. The chef who caves to the restaurant owner may have saved their job for a few weeks, but they've also set themselves on a different path. They'll move from shady restaurant to shady restaurant, always working for one creep or another.

Be the person you'd want cooking your own meals. People are the hardest design problem you will ever face. And there's no pattern for solving them, because people are messy.

THE MORE YOU NEED MY HELP THE MORE I LOVE YOU

I decided to be a designer because I enjoyed designing things. I enjoyed designing things because it helped people. Go be helpers. It's a really awesome thing to do.

So yeah, I love design, and how it improves people's lives. I love people who dedicate their lives to improving the lives of others. And in an ethical design world, that's what we're all doing: working to improve the lives of others. Starting with those who need the most help. And if that all sounds a little too kumbaya, well fuck it—this is the happiest article you will ever read about design ethics! So we're gonna get a little kumbaya.

Are there going to be obstacles put in our way? If there weren't, we wouldn't be getting paid to solve this stuff. Are those obstacles sometimes put in place by the very people paying us to remove them? Yes. They're paying us to do a thing because we're good at a thing and they're not, so we can't expect them to do things the same way we would. (Sometimes people stand in the middle of the kitchen because they think it's helpful. Let them feel helpful—tell them to go take drink orders.)

At the end of a good meal, if you were raised right, you'll tell your waiter to give your compliments to the chef. I don't doubt even the most renowned chefs enjoy hearing that. No matter how long you've been doing something, it always feels good to hear your work had a positive effect on someone's day. You did the work. You put in the effort. You sweated the details. You used the best ingredients. And you get acknowledged for it. But that's not gonna happen for designers. No one finishes an online banking transaction and says "Please send my compliments to the design team." But maybe it would make a difference, so let me be the one to do it: my compliments to the design team. Thanks for taking care of the person on the other side of the screen. They deserved your best work, and they got it.

Kumbaya.

DOING THE RIGHT THING THE WRONG WAY

(Originally published in Dear Design Student on December 21, 2015)

> *Q: You're always going on (and on and on) about how designers are responsible for what they make. Do you think everything out there is a bad idea?*

Picture the following scenario: You're downtown and you're trying to get to an appointment. Say a medical appointment. Your doctor's office is in a part of town that isn't served very well by public transportation. So, you hail a cab. Except the city doesn't have a lot of cabs. You call dispatch and they say they're sending one. It never shows up. This is frustrating. You're now worried about missing your appointment.

Now, what if we could solve this problem by putting more cars on the street? And say we could turn private citizens into drivers? You've not only solved the dearth of cabs problem, but you've found a way for people to make a little extra scratch, possibly even earn a living by driving other people around.

All you need is a way to connect the people who need rides to the people who offer rides. That's actually the easy part. You just build an app. (Well, maybe not easy, but you get what I'm saying.)

By all accounts, this is a good idea. You've found a problem. You've come up with a reasonable solution. You've met demand with supply. You've found a way for people to earn a living, including yourself, hopefully. Certainly, there will be hurdles along the way. The cab companies might not be happy for one. But there is nothing ethically wrong with this idea.

Let's look at another scenario: You're going out of town on business. You attempt to book a hotel. There's a conference in the town you're going to, so prices are jacked up. Now you're annoyed because you're paying more than you wanted to. Certainly more than a room is actually worth. And you hate staying at hotels anyway. They're sterile. The air sucks. The shower controls are complicated. You can't open the windows. At the last minute, you remember a college buddy lives in that town. You call him and he invites you to stay with his family. You have a wonderful time catching up.

And it occurs to you that there are millions of people out there who might have a spare room, like your buddy. You wonder if they might want to rent those rooms out to travellers. That way travellers could feel at home when visiting a foreign city. You've solved the bad hotel problem, and you've figured out a way for people to make a little extra scratch.

All you need is a way to connect the people who need rooms to the people who offer rooms. That's actually the easy part. You just build an app. (Well, maybe not easy, but you get what I'm saying.)

By all accounts, this is also a good idea. You've found a problem. You've solved it. You've met supply with demand. You've found a way for people to earn a living, including yourself, hopefully. Yep, there'll be hurdles. The hotel companies might not be happy for one, but again, there is nothing ethically wrong with this idea either.

Obviously, we are talking about Uber and Airbnb. Two massive success stories. Also, two companies that have been vilified by many in the past (including by me). Two companies that seem to be constantly at odds with the cities in which they attempt to do business. Two companies which various cities have tried to ban. Two companies, each with numerous cases of safety issues, harassment of customers, inadequate customer service. Two companies with death tolls. And two companies with leadership teams who spend an inordinate amount of time with their feet in their mouths.

So how do ideas, which start out helpful and by all measure ethically sound, turn into companies with the ethical charm of a decapitated horse head bleeding out onto your silk sheets? Easy. You introduce people. Even easier, you introduce people with a very narrow set of life experiences.

The scenarios described above are utopian. They work great as long as everyone behaves well. And by everyone, I mean everyone from the company founder to the person providing the service to the person using the service. But as Chekov once (maybe) said, if you introduce a person in act one they'll probably turn into an asshole by act three. Services which rely on people are guaranteed to have assholes at every level of the supply chain. And while the ideas themselves may not be unethical, the execution of those ideas will have ample opportunity to turn up unethical and clueless designers.

Especially when those designers all have the same life experiences. Celebrate the same holidays. Went to the same school. Look like each other. In other words, white boys solving problems for white boys. They've never been harassed, so they don't think of solving for that problem. And even if they do, they don't solve that problem from a place of experience. They've never had a cab refuse to stop for them, so they don't solve for that problem. They've never had a host refuse to rent them a room based on race, so they don't solve for that problem. They've never had a host be a little too eager to rent them a room, so they don't solve for that problem. And it's too easy to think that terrible things don't actually happen as often as they happen. But they do.

We owe it to the people we are designing for to build our teams to reflect those people. Don't assume how a woman would behave in that situation. Get women to design it! Don't assume how a Black person would behave in a situation. Get Black people to design it! Empathy isn't enough. We need inclusion.

The point isn't that any particular experience or classification makes you a better designer. People are more informed about themselves than about others. And for now the vast majority of people in tech are white males. (By the way, current projections have whites becoming a minority in the United States around 2040. I look forward to their call for minority inclusion right around then.)

Hire people who have been dealing with assholes their whole life.

Earlier this evening, a good friend shared a link with me to Michael Moore's "We are all Muslim" campaign. Now, I'm a big fan of Michael Moore and his great big socialist heart. I'm also a big fan of my friend, he was sharing the link with the best of intentions. But Michael Moore is wrong on this one. We are not all Muslim. We can empathize. We can pretend. But at the end of the day, we will not have a presidential candidate calling to banish us from our country because of our beliefs. This country is not a melting pot. This country is a thousand cultures. All living together. Hopefully in harmony, but not so much these days. And it's those differences that we need to celebrate. Not the sameness. The sameness is boring.

And as the products and services we build get more and more enmeshed in that weird-ass complicated social fabric, our teams need more and more, to reflect those differences. A diverse team isn't just about the diversity of race or gender. It's about the diversity of experiences, diversity of needs, diversity of thinking, and ultimately diversity of solutions.

Diversity is our strength. We're idiots for not using it.

And yes, design, when done right, is always political.

BEWARE THE JUDAS GOAT

(Originally published in Modus on October 4, 2019)

> Q: A few weeks ago something bad happened at work. It upset me enough that I discussed it with HR. They said they'd handle it, but they never got back to me about it, and the bad situation is still happening. Should I follow up? Should I forget about it? Should I trust that HR handled it?

Let's talk about sheep. Besides being a handy metaphor, sheep are herd animals. They enjoy each others' company, because it not only provides safety, but it also gives them other sheep to hang out with. It's a community. The group names for sheep are flock, herd, and mob, words that are also used to describe groups of people in various contexts. Maybe that's why they're a handy metaphor. Another reason might be because they're highly suggestable and very trusting. It's easy to get a sheep on your side. Especially if the sheep thinks you're one of them.

Enter the Judas goat.

Judas goats are used by ranchers to herd goats from pasture to pasture, and eventually to slaughter. The goat is trained by the ranchers to follow their commands, then gets in with the sheep, who accept the goat as one of their own. In time, the goat becomes a trusted member of the sheep community. Probably because they show up with donuts on Friday. The sheep feel comfortable around the goat. They tell the goat their problems. They tell the goat when the other sheep are treating them badly. The goat becomes not just one of them, but someone they feel they can count on. This is part of the rancher's plan. On the day of reckoning the rancher signals the Judas goat to gather up the sheep and bring them down to the slaughterhouse, which the goat is more than happy to do, because its own life is spared. The Judas goat never suffers the same fate as the sheep. It lives to be introduced into the next herd. Once trained, the goat is valuable.

THE COMPANY JUDAS GOAT

A Judas goat understands it works for the rancher, much like an HR department understands it works for the company. It exists to protect the company's interests, not yours. But to protect the company's interests in the best way possible, the HR department needs you to think they work for you. They mingle among you. They bring snacks. They invite you to come and talk to them if you're having a problem with your job, or with a coworker, or with management. Their ultimate goal is to get you to think they're on your side. They are not.

"But Mike, I know some really good HR people!"[1]

I'm sure you do; I know some really nice goats. No doubt there are people in HR who actually believe they're on the employees' side. For all we know, the Judas goat thinks it's taking the sheep for a nice walk. The biggest issue here isn't individual character, but the design of the system. HR is hired by management, paid by management, and ultimately reports to management. They are company officers. They may even be the most well-meaning company officers. But ultimately their allegiance is to the company. Or as Brad, the CEO, may have to remind them every once in a while—"to the big picture." No sheep is bigger than the ranch. Oh, and they probably own some company stock as well.

When Susan Fowler bravely described her harassment while working at Uber, HR was very much a key player in the story. Her boss started sending her harassing chat messages during her first week at work, and as she wrote in her blog (and in an upcoming book) she immediately took those messages to HR. Her expectation was that HR would "would handle the situation appropriately, and then life would go on." That's a reasonable expectation. As workers, we're trained to do this. And we expect that the problem will get handled by the people who are paid to handle the problem. In actuality, and as Susan Fowler discovered, that's not the case:

"When I reported the situation, I was told by both HR and upper management that even though this was clearly sexual harassment and he was propositioning me, it was this man's first offense, and that they wouldn't feel comfortable giving him anything other than a warning and a stern talking-to."

We now know this wasn't the man's first offense. There'd been several previous offenses, and they were reported to HR as well. But as Mike Isaac details in his book *Super Pumped*, the man was

---

1 See also good cops, good priests, and good Republicans. Individual goodness doesn't negate systemic corruption.

what Uber called a "high performer." His harassment was tolerated by the company because he was earning. HR wasn't protecting the victim of harassment. They were protecting the perpetrator, whom they saw as a valuable asset. I'd also point out that in the quote above, HR and upper management spoke with one voice. Clearly they had discussed the problem together and gotten their messaging in sync.

HR works for management. The Judas goat works for the rancher.

JUDAS GOATS INSIDE US

The Judas goat is bigger than one person or one department, though. Ultimately, we're all our own worst Judas goats. The imposter syndrome. The nagging doubts that we're just lucky to have a job. The belief that we haven't earned the right to be free. The emails we answer at 11pm. The Saturdays we give up. The feeling that maybe we did something to bring on someone else's bad behavior. The belief that if we eat enough shit and keep our heads down we'll somehow gain our abusers' respect.

In their excellent time travel novel, *The Future of Another Timeline*, Annalee Newitz points out that sometimes to correct a problem in the past, you have to travel further into the past than you were expecting to. For example, if you wanted to correct the 2016 presidential election, you'd want to travel back to 1812, when the first case of gerrymandering was reported by the Boston Gazette, or even to 1787 when the Three-Fifths Compromise was struck, or... you get the idea.

To figure out when we became our own Judas goats requires us to go back to when we gave up our rights as workers. Rights that people died for. The 40-hour week. Weekends. Collective bargaining. Overtime. Benefits. The right to organize and meet management on a level playing field as equals.

To paraphrase John Steinbeck, tech workers never saw themselves as workers because they saw themselves as temporarily embarrassed founders, entrepreneurs, or C-suite executives. Being a worker was a short-term stint. We traded rights for stock options. We traded collective well-being for a future that—and I hate to break this to you—most of us will never see.

We gave away all our worker's rights without any thought to what we were doing to the workers that came after us or the communities it would destroy. We Judas-goated our own descendents.

We've hijacked workers' communities. When you live in company housing and wear the company clothes and eat in the company cafeteria and spend your free time in company outings and get health care from company therapists, that company is your community. And you protect that community at all costs.

A herd that leads itself willingly to slaughter doesn't need a Judas goat. It contains multitudes of them.

I once sat in a café in San Francisco and overheard a young man, presumably a tech worker, complain to his friend that he was concerned he wouldn't make his first million before he turned 30. This. Is. Not. Normal.

And ultimately, it's not healthy. But even worse, it's selfish. Our industry promotes the accumulation of vast individual wealth by squeezing those who can least afford to be squeezed. We applaud initiatives that reduce or even completely eliminate workforces to maximize profits. Uber, who built its wealth on the back of a labor force it won't even recognize, is now pouring those profits into the elimination of that same labor force by putting their eggs in the self-driving car basket.

A GOAT OF OUR OWN

The interface to management cannot be controlled by management. The voice of the workers must come from the workers. Issues with management cannot be filed with a representative of management. They need to be taken to a body which represents the workers. One which cares for the workers. One which is made up of the workers.

Imagine if Susan Fowler, and others in her situation, of which there are many, didn't have to take their issues to the company Judas goat. Imagine if a worker who was afraid of losing their job didn't feel like they were fighting alone. Imagine if a worker who was let go had a path to file a grievance with a body made up of other workers like them. Imagine if a worker who was asked to do unethical work felt empowered to say no because they knew they had the support of an organization who represented their best interests, not the company's.

Last week Kickstarter fired two employees who were trying to organize their workforce. Two people lost their jobs for attempting to improve their working conditions. That's unacceptable. There was a time when an employee could take their issues to a union representative. Management dealt with employees as a herd, a flock, and yes—sometimes a mob, when a mob was called for.

We have forgotten what those times were like. It's time to remember.[2]

---

2 *If you're union-curious, the CWA (Communications Workers of America) has started an initiative to unionize tech and gaming workers. They're good people, and they will look out for you and yours. Take a look at some of the success stories here: www.code-cwa.org/*

THE BEST THING WE CAN DO FOR THIS PLANET IS DIE

(Originally published on Medium on March 20, 2018)

A few weeks ago, I was in Copenhagen giving a new talk. I get nervous with new talks, not because public speaking makes me nervous, but because you never know whether a new talk sucks or not until you've given it a couple of times. And you don't even really know what the talk is about until you've given it a few times. And it wasn't until I was in the middle of this talk, which was ostensibly about ethics, that I realized it had a strong undercurrent of death throughout. And maybe undercurrent isn't the right word. It's quite possible that if you asked someone in the audience what the talk was about they would've replied "Death. That was some dark shit."

Death rules everything around me.

Later that evening I went out to dinner with a couple of friends. We went to a place that specialized in "Nordic", which I assumed meant eating whale and drinking mead while *Thor Ragnarok* played on a giant screen above the bar. But it ended up being a very nice cozy place, with an even nicer owner. The kind of guy who grabs a bottle of bourbon, pulls up a chair for himself, and proceeds to tell you about spending ten years in the Danish military. In between stories of Finns building saunas in Kabul, the topic of "being good allies" came up. To which my new Danish friend shouted that men our age had committed too many sins and done too many things wrong to ever be good allies in any sense of the world. And the best thing we could do for the planet was to die.

Last week I watched as American school children walked out of school in protest. Because they're tired of going to school and getting shot. Because they're tired of their government caring more about fleecing their own pockets than comprehensive gun control. Because they're tired of their classes being interrupted to practice active shooter drills. And more than a few of them are tired of actually burying their classmates. And as I'm watching these brave brave kids, I'm filled with equal amounts of hope for the courage they're displaying and shame that our generation has left this problem for them to solve. We're a year away from the 20th anniversary of the Columbine High School massacre. We should've taken care of it then and there, before these brave kids were even born.

I start thinking that maybe my Danish ex-military friend is right. The best thing we can do for this planet is die. Death rules everything around me.

Societies are not made up of laws as much as it's made up of an agreement to follow those laws. And while laws are delivered to us in a top down fashion, the agreement to follow those laws is upheld from the bottom up. A code of ethics will not magically transform us into people who behave decently. It's imposition, coming from the top, will have no transformative power. Only an agreement to follow it, made at the rank and file level, can change how we work.

This is where my hope comes from. I believe the people coming up after us will do a better job than we did. I believe that as a 50 year old white male living in America, my goal is to clear the path for the voices I've silenced either knowingly or unknowingly. I cannot be a good ally because I've benefitted too much from the world I was born into. And regardless of whether I wanted those benefits or not, I got them.

> "If you are white in a white supremacist society, you are racist. If you are male in a patriarchy, you are sexist."
> —Ijeoma Oluo, So You Want to Talk About Race

As uncomfortable as it is to admit, I have benefitted from both racism and sexism. And if you are reading this and you look like me, you have too. Regardless of how well you've lived your life, regardless of how good your intentions were, you benefited from a stacked deck. And yet, even with the deck stacked in our favor, we couldn't do the job. So yes, the best thing we can do for the planet is to die.

Death is always a given. It is not a choice. As a culture, we spend a lot of time attempting to delay it, or comically convincing ourselves it's not coming. But there's absolutely nothing we can do to stop it. It. Is. Coming.

And rather than spend a lifetime convincing ourselves that it's not, and wasting our energy attempting to outrun it, perhaps we are better served in attempting to earn it. Perhaps, just perhaps, the point of life is to earn the death that comes at the end. And perhaps, no most likely— that death is best earned by doing everything we can for those coming up after us. Earn your death by making room for the generation behind you. Might they fuck it up as well? Of course. But you already have. They still have a chance.

Death rules everything around me. Let's do at least one thing right. Let's die well.

THE ROAD BACK
What we need to do to get this industry back on track

(Originally published in Modus on December 11, 2019)

> *"Will you please just write something positive before this year is out?"* —My Editor

The promise of the internet was that it was going to give voice to the voiceless, visibility to the invisible, and power to the powerless. That's what originally excited me about it. And I'm guessing that's what originally excited a ton of people about it. It was supposed to be an engine of equality and democracy. Suddenly, everyone could tell their story. Suddenly, everyone could sing their song. Suddenly, that one weird kid in Trondheim, Norway, could find another weird kid just like them in Bakersfield, California, and they could talk and know they weren't alone. Suddenly, we didn't need anybody's permission to publish. We put our stories and songs and messages and artwork where the world could find them. For a while it was beautiful, it was messy, and it was punk as fuck. We all rolled up our sleeves and helped to build it.

And then suddenly, we broke it.

This was the web in 2009: Social was taking off, Shaq was tweeting free game tickets, we were sharing jokes, finding old friends we hadn't talked to since grade school, and making new ones. There was optimism in the air. Money was free-flowing. (Tech) jobs were free-flowing. We were excited that collecting all this data on people would lead to amazing insights about something. (We just weren't sure what.) And we thought we'd just helped elect America's first internet president.

This is the web in 2019: toxic anger, hate, actual motherfucking nazis(!), stolen data, gender reveal parties, monstrously large corporations behaving monstrously badly, and America's first actual internet president—willfully allowed to rise to power because the web is ruled by engagement and run by idiots who wrapped themselves in free speech, while not understanding what it meant. Fascists may have rolled into town, but they rolled in on roads built by libertarians.

And meanwhile, we also broke the planet.

So, is there a road back? After all, I aim to make my editor happy and she has asked for something positive and I am aiming to deliver. Do we have a shot? We may. We may, on the most optimistic of days, have the smallest sliver of a shot. If we aim correctly, and blot out all the anxiety and despair surrounding us, we may even be able to see it. A small glimmer of hope, thin as a thread, but it's there.

I recently read Jonathan Safran Foer's We Are the Weather, which posits the same question about climate change. Do we have a chance? (I encourage you to read it.) His answer is much the same; we have a definite maybe. But only, only, if we start right now, agree on the right action to take, and if we don't hesitate. And honestly, that doesn't sound like a good chance. We've only rarely been able to do any of those things. And all three combined seems impossible. In We Are the Weather, Safran Foer tells a story of a woman in a car accident who manages to lift the car to save her child trapped under it. She exhibits superhuman strength simply because not saving the child is impossible to consider. When failure becomes impossible, success, no matter how unlikely, becomes the only possibility.

We don't have much of a chance here, but not succeeding is impossible to consider. So we have to consider what success might look like, as it's the only option left. So grab that small sliver of a shot and hang on with all the strength you don't realize you have yet. Because you do have it. It won't be easy. You won't like it. But holy shit, if it works?

Suddenly, we've fixed it.

TAKE RESPONSIBILITY

The world was broken on our watch. Non-negotiable. Because we don't have time to negotiate the point. We were given the responsibility to use our labor and our expertise to make the world a better place, and we failed. We have, in fact, made it worse. You are forever responsible for the work your labor produces. And if you're not okay with that I implore you to stop producing work.

On December 3rd, ProPublica published a story about the ongoing work McKinsey Consulting has been doing for ICE. It included recommendations that, apparently, were insane enough to make people who keep babies in cages uncomfortable:

> ...the money-saving recommendations the consultants came up with made some career ICE staff uncomfortable. They proposed cuts in spending on food for migrants, as well as on medical care.

A few days later, McKinsey published a rebuttal that basically said the story wasn't true. I've worked with ProPublica. They're passionately committed to the truth. They wouldn't have published a word of that story if they weren't absolutely certain of it. And McKinsey reacted the way organizations like McKinsey react. Deny everything and wait for the next news cycle. It's a good strategy. The next news cycle will inevitably contain news of another company doing something even worse. Probably Facebook.

But here's the thing: Every employee at McKinsey who worked on this knows they worked on it. Every employee at McKinsey is complicit in this lie. And every employee at McKinsey has an opportunity to stand up and say, "No. We're lying. We did this."

If we need to get back on track, we need to take responsibility for our work. Take your shot, McKinsey employees.

BUILD WHOLE TEAMS

For the last ten years, we've been focused on making teams leaner and getting them to work faster. Projects were carved into tasks, and tasks were carved into stories, and everyone got really, really good at making their little part of the whole while having no idea what the whole was, or how the whole affected society. Because the real goal here was to get this shit out the door and get to our venture capital-mandated exit event before everyone realized there was no business model, path to profitability, or actual service here.

In her excellent book, Broadband, Claire L. Evans tells the story of the venerable Grace Hopper working at the Harvard Computation Lab during World War II, tasked by the Navy to figure out a complex partial differential equation. As Evans tells us, Grace Hopper had absolutely no idea what the equation was for; to her it was just an interesting mathematical challenge.

The partial differential equation turned out to be a mathematical model for the central implosion of the atomic bomb. Grace never knew, until the bombs fell on Nagasaki and Hiroshima, precisely what she had helped to calculate.

Would Grace Hopper still have solved the equation had she

known what it was being used for? We'll never know. That choice was taken from her. And she deserved the opportunity to make that choice. Just as we all deserve the opportunity to research and understand the possible ramifications of our work.

Will being circumspect slow us down? Absolutely. But with apologies to the venerable Grace Hopper, that's not a bug.[1] That's a feature.

UNDERSTAND YOUR POWER

This job isn't about helping Nike sell shoes. It's about making sure everyone has shoes.

This job isn't about creating bullhorns for fascists and others who'd use their power to denigrate others. It's about smashing the fascists' bullhorns.

This job isn't about building tools that hand our data to the corporations of Silicon Valley. It's about building tools to keep that data away from them.

This job isn't about creating shareholder value. It's about creating human value.

The job isn't about making rich white men richer at the expense of everyone else. It's about making sure everyone earns an equal share for their labor.

But how will we pay our rent, you ask. Tech workers are some of the highest paid workers on the planet. You're not Jean Valjean. You're not a loaf of bread away from your family dying of hunger. Your Peloton subscription is not worth putting a child in a cage. Your argument is invalid.

For too long, we've treated the job as if we were servants. We did what we were told. We followed orders. We didn't ask questions. We may have rolled our eyes once in a while when something didn't seem right, but we did it anyway. We behaved as if we had no say and no agency in how the job was done. We lost control of our labor, our hands, and finally our minds.

Yes, design is political. Because design is labor, and your labor is political. Where you choose to expend your labor is a political act. Whom you choose to expend it for is a political act. Whom we omit from those solutions is a political act. Finally, how we choose to leverage our collective power is the biggest political action we can take. If we choose to work collectively, we have a

---

1 Grace Hopper logged the first computer bug. And, yes, it was an actual bug. In Harvard's Mark II in 1947. Look it up. She was an amazing pioneer and I feel kinda bad that I highlighted this particular story about her, because she broke a ton of barriers and was pretty badass. You can find out a lot more about it in Claire L. Evan's Broadband. It's a really good book.

ton of power. If we continue to behave like servants, we're not just letting ourselves down, we're letting down everyone whose lives we swore to improve.

We're late to the party. The world is working exactly as we designed it to work, and that's the problem. We're here because we've abdicated our responsibility. We're here because we forgot how much strength we have when we act together.

It's time to remember who we are. It is time to unionize.[2]

RESPECT YOUR TRUE COMMUNITY

Twitter employees love posting pictures of events at the workplace, and they seem to have a lot of events. Those photos are usually accompanied with the hashtag #lovewhereyouwork, which I'm sure is something the company encourages. Both the events and the hashtag are meant to inspire a sense of workplace community, as well as to show the world how happy all the employees are.

(Stay away, union organizers, you shall find no purchase here!)

Same with the free meals, the offsites, the kombucha on tap, the homey surroundings, the on-site perks, etc. Love where you work. And try to forget that we're using your labor to line our shareholders' pockets while destroying society.

Obviously, this isn't limited to Twitter. Tech companies are known for their sprawling luxurious campuses with lots of perks. People need community. We're herd animals. We like to surround ourselves with others, and preferably others with whom we share common interests, causes, and backgrounds. (This isn't always a good thing. Sometimes we define our communities by skin color and by who it excludes.) But communities tend to look out for each other and to circle the wagons when the community is in trouble.

Tech companies have hijacked the notion of community, and they've done this by design. Most tech workers are first and foremost members of their company community. The company looks out for them, and they protect the company's interest.

Facebook was even nice enough to build a chatbot for their employees to help them deal with their families' difficult questions about the company during the Thanksgiving holidays. The only other organization that I can think of that reaches that level of "family management" is Scientology.

Stop loving where you work. Love where you live. And don't live at work.

---

2 *Once again, for the people in the back: www.code-cwa.org/*

IT'S NOT MADE BY GREAT MEN

The first barber shop opened to the public in 296 BCE. I know this because I googled it. I googled it because I was looking for a metaphor. Specifically, I was looking for a metaphor about people who were good at what they did, and then decided that being good at that one thing meant they must be good at everything else. Barber shops immediately came to mind because, as some of you may know from trivia night, for a time you could get dental work done at the barber shop. As well as some surgery. Turns out the reason for this is amazingly idiotic. Barbers already had the tools and they already had you in the chair. As long as you're getting a haircut and a shave, we might as well get that impacted molar and do a little leech work.

It's also a good metaphor to use when talking about the tech industry, because much like the tech industry, barber shops have historically been the purview of men and were places where you could find a lot of porn.

Mark Zuckerberg is good at coding. At least I think he is. The last thing I know he actually coded was a website for rating co-eds, which turned into Facebook. I'm sure Jack Dorsey is probably good at something too. But both these fools are terrible at interpersonal relationships, which makes it appalling that they're both in charge of platforms which rely on understanding how human beings relate to each other. They've both proven that's not in their wheelhouse. And just like none of us would let our barbers do our dental work, or let our pets do our tax work, it's time to understand that the challenge of the web is no longer technical. And that being good at the technical stuff doesn't automagically make you a savant in socio-economic policy. The web is about people and how they interact with each other. The web is about power dynamics. The web is about society and its discontents.

The next wave of people running the web need to understand human relationships more than they need any other set of skills.

GET OUT FROM UNDER VENTURE CAPITAL

I'm writing this in a café in San Francisco. There are at least three adjacent tables of people talking about funding. I'm not eavesdropping. The conversations are meant to be heard. They're a social mating call. Look at me, I'm a founder, raising capital. Look at me, I'm an angel investor, finding the next Uber, but for cats.

Look at me, I'm a freelance writer, desperately trying to write something positive for my editor.

It was at this same café where I once heard a young guy complaining to another young guy that he was worried he'd turn thirty before he made his first million. This isn't normal. This isn't sustainable. And, more importantly, this isn't good.

For generation upon generation, people built and ran businesses. Trying to make more than they spent. Trying to increase that number a little bit year after year. All the while, attempting to adapt to customers' ever-changing needs and desires. Multiply that with a whole lot of luck, and divide it by a whole lot of racist and sexist bullshit about who could get business loans and leases. But in a nutshell, that's how we built businesses. And the entire goal was to make sure the generation that was coming up after you got a chance to do a little bit better than you did. Rinse and repeat. But you built things a step at a time, with the steady stubbornness and surefootedness of a good working mule.

Now suddenly, we have Sand Hill Road.

But the internet was a whole new industry, and it was exciting. And exciting things grow quickly. The money showed up fast. Suddenly, every boy in a hoodie had the next potential great idea, and if you could get a good grip on his (always "his") shorthairs early and right by the base, you stood a chance to get ten times richer. So, the investments came fast and easy. And suddenly these cuddled little runts were managing millions of dollars and thousands of people and hundreds of problems. With absolutely no idea how. In the world of venture capital, this doesn't really matter, though. As long as we push the hog into the slaughter chute before the inspector realizes it's riddled with worms, we still get paid for the meat.

Adam Neumann, formerly of WeWork, broke his company, broke his workforce, and was given $1.7 billion dollars to walk away. Travis Kalanick, formerly of Uber, a company responsible for 3,045 sexual assaults, 9 murders, and 58 people killed in crashes in the last year alone[3], was allowed to walk away with almost $3 billion. Jack Dorsey, who built a $4 billion dollar fortune by enabling Donald Trump and the alt-right to use his platform for spreading abuse and hatred, now intends to spend the 2020 election yogababbling, fasting, and downward dog-whistling in Africa. And Mark Zuckerberg turned a website for rating the hotness of college women into an engine for destroying democracy and our privacy and a $74 billion bank account.

And, lo, we cannot completely put this at these idiots' feet.

---

3 *None of this is hyperbole, by the way. These are actual numbers. Check out Mike Isaac's book about Uber: Super Pumped: The Battle for Uber, which should be in the True Crime section of your local bookstore, but is probably actually in Business. Who am I kidding, though. You're gonna Prime it.*

The Righteous Angers

As Professor Scott Galloway recently said in an article in Business Insider:

> *...if you tell a thirty- or forty-something person, who regularly wears black turtlenecks, that they are Steve Jobs, they are inclined to believe you.*

The venture capitalists who raised these sick hogs did even better financially. And in the end, none of these services exist as anything but a means to make their investors richer. All of them are a stain on society. And that's okay, because in the end the only thing any of those companies were actually expected to do was to make their investors rich.

Again, Professor Scott Galloway says:

> *It's not Mr. Dorsey's plans to move to Africa that constrain stakeholder value, but his plans to move back. Mr. Dorsey demonstrates a lack of self-awareness, indifference, and yogababble that have hamstrung stakeholder value.*

Suddenly, guillotines.

No matter how great your idea is, once you take venture money your goal changes from developing that idea, to making sure your investors get their payday. I want to see you develop your ideas. And I want to see you develop ideas that are good enough to survive in the world, as well as improve that world.

Maybe you didn't make your first million before you turned 30, but that wasn't why you showed up, was it? And if it was, let me show you the door.

...

The promise of the internet was that it was going to give voice to the voiceless, visibility to the invisible, and power to the powerless. That's what originally excited me about it. That's what I hope excited you about it too. And I hope that some of that excitement is still there. But as we've learned, hope, by itself, is worthless. We need a plan. And we need it soon.

We don't really have time to argue about it. The things we can still save are very much worth saving. It would be unthinkable not to save them. It will be hard. It will be close to impossible. But we can't fail. When failure becomes impossible, success, no matter how unlikely, becomes the only possibility.

We wanted to change the world. We have just enough time to take one shot. Let's make it count.

PART TWO

The Tech Bro Angers

Gilly & Billy drawing by Adam Koford

> "People have the power to redeem the work of fools."
>
> — Patti Smith

ONE PERSON'S HISTORY OF TWITTER, FROM BEGINNING TO END

(Originally published on Medium on October 15, 2017)

At some point in 2006, or possibly late 2005, Noah Glass[1] walked into our office all excited about something. That in itself isn't news because Noah was always excited about something. Dude had an energy. Noah worked across the hall from us on the sixth floor of an old, broke-ass building in South Park. He came over all the time. He was friendly like that. Here's why we're talking about this particular visit: Noah was excited to tell us about a new thing he was working on. "You can use it to send group SMS."

"That sounds stupid."

"Look at the logo!"

"That's even stupider."

This was my first look at Twitter, or twttr, as it was annoyingly called then. I was right about the logo, and wrong about the service. It wasn't stupid, it was just hard to explain. So, Noah showed it to me, and I still thought it was stupid. I'll admit that it took me a while to get it. I didn't care what people were having for lunch. I didn't care where people were at. (Remember, this was also the golden age of check-in services, where people made sure all their friends knew where they were at all times lest they subject themselves to a moment of introspection.) Nevertheless, I signed up anyway, tweeted a few times, and was fairly close to deleting it a few times as well. Until one morning, I was in a cab headed to therapy, which meant I was in a mood, and I absent-mindedly tweeted out "I've been shot!" then turned my phone off and went to talk to my therapist about becoming a well-adjusted human being.

---

[1] If you're wondering "who the fuck is Noah Glass?" I wouldn't blame you. Not since the days of Stalin has someone been so completely erased from history. Noah was Twitter's co-founder, and arguably its driving force, in the very early days.

When I turned my phone back on I had about 20 new messages. Texts, voicemails, and a bunch of tweet replies. Including my now-wife, wondering what hospital I was at. That's the day I discovered what Twitter was for. It was for having fun. And telling jokes. (BTW, my wife still doesn't think this was a good joke.) That's when I was hooked.

The first few years of Twitter were fun. The jokes piled up. I met other people who also liked to tell jokes. We even told a few non-jokes once in a while, like when someone was going through a hard time, we'd stop telling jokes long enough to make sure they were okay. Then we'd go back to jokes. But seriously, it was mostly about making jokes. We even had a website that turned our stupid jokes into a deadly leaderboard game:

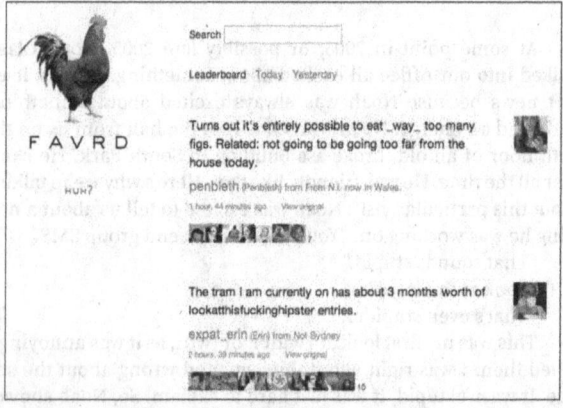

At the same time we were telling these dumb jokes, we were meeting a lot of new people. Twitter was a really great way to meet people. And believe it or not, there was a time when you could say hello to someone and "fuck you, you lib cuck" wasn't the first reply that came to people's minds! Some of my very closest friends today are people I met during the early days of Twitter. And here's the really amazing thing: I still haven't met some of those people. And some of them live halfway around the world.

I also know people who met on Twitter and have children now! Well-adjusted children. This happened. People met. Tweeted at each other. Got little crushes on each other. Figured out ways to meet in person and then made babies! People met on there and started businesses. Advertised jobs. Found jobs. Celebrated kids' birthdays, and got support when a loved one died. Then inline images happened and we added photos of cats, kids, and sad Keanu to this wonderful mess.

Twitter also taught me how to be a better writer. (Count how many of these sentences are under 140 characters.)[2] Seriously. I'm actually a pretty introverted person, and Twitter was a great way to shake that. (I wanted to shake it.) But as stupid as this might sound, every little star (they will always be stars) gave me a little more confidence. And eventually, what started as a place to tell jokes became a place to talk about design. And I got confident enough to start sharing those ideas too. I've written two books about design, and I can trace both of their origins to shit I said on Twitter. And when I was writing those books, I kept my book in one window, and Twitter in another window. If I thought a sentence was pretty good I pasted it into the Twitter text field to make sure it was 140 characters. Twitter made me a better writer.

My first editor is probably reading that line and nodding and thinking "Fuck you. I made you a better writer, asshole." And that's true. But I met her on Twitter.

There was a time when Twitter was a place you went to fuck around, and accidentally made friends and got smarter. It's been years since I've felt smarter after being exposed to Twitter, but trust me, those days were real. They happened.

...

I moved to San Francisco in 1999. I was part of that wave of people who moved here and destroyed the city. But I didn't come here to get rich. (Every subsequent decision I've made since then bears this out.) I came here because the internet was new. It was exciting. It felt like punk rock for publishing on a global scale. We could make things and just tell the world. We didn't need permission. We were gonna build a new way of communicating and everyone was gonna have a voice in this new world.

I wasn't the only idiot who felt this way. San Francisco was full of hope and crazy back then. And sure, a lot of it was driven by money, but more accurately, it was driven by giving young hopeful kids a lot of money and waiting for them to make something that could make you money. They were exciting times. We were young, stupid, and equipped with more processing power than any human being had ever possessed in history. (Also, a lot of fucking ecstasy.[3]) I remember walking around the city on those days. Lotta hope. Feeling a bit cocky to be honest. But we thought we were

---

2 *This was written when the character count was still 140. By the way, Twitter expanded the character count to 280 at Donald Trump's behest. You can't prove this isn't true.*
3 *I think kids call it molly now.*

gonna change the world.

Here's the bad news: we did.

Twitter was built at the tail end of that era. Their goal was giving everyone a voice. They were so obsessed with giving everyone a voice that they never stopped to wonder what would happen when everyone got one. And they never asked themselves what they meant by everyone. That's Twitter's original sin. Like Oppenheimer, Twitter was so obsessed with splitting the atom, they never stopped to think what we'd do with it.

Twitter, which was conceived and built by a room of privileged white boys (some of them my friends!), never considered the possibility that they were building a bomb. To this day, Jack Dorsey doesn't realize the size of the bomb he's sitting on. Or if he does, he believes it's metaphorical. It's not. He is utterly unprepared for the burden he's found himself under.

The power of Oppenheimer-wide destruction is in the hands of entitled men-children, cuddled runts, who aim not to enhance human communication, but to build themselves a digital substitute for physical contact with members of the species who were unlike them. And it should scare you.

On November 8, 2008, I watched Barack Obama win the presidency of the United States while I was sitting on Twitter's office couch. I forget who invited me, but I was excited to be there because this felt like the first presidential election that the internet had an active part in. Whatever that meant. It felt like all of the tools the web community had spent the last ten years or more building had actually culminated in this moment. And I sat on that couch crying. I was getting to see this moment as a guest in the place that got all of these voices communicating. And all of those voices helped elect a president. In 2008, I thought Twitter helped elect a president. I was off by eight years.

. . .

Alas, helping to elect presidents and taking credit for global movements isn't enough to ensure company growth. From its inception, like most startups from the era, Twitter lacked a clearly defined business plan. Turns out changing the world isn't a business plan. Now I am not a business expert. In fact, I know jack shit about business plans.[4] So I won't go into details about how or why or whatever else because I'd be making it up, and the internet is full of hot takes on this already. Some of them are written by smart people.

---

4 *This isn't completely true, by the way.*

I will say this though: the goal of every venture-backed company is to increase usage by some metric end over end over end until the people who gave you that startup capital get their payday. This is the original sin of Silicon Valley. And Twitter had plateaued, and in the Valley plateauing is a thousand times worse than flaming out. Twitter needed a spark. Twitter, not realizing they were sitting on a bomb, went looking for something to light the fuse. They were about to get it.

...

In March 2009, Donald Trump joined Twitter. No one noticed. Why would we? He was a washed-up NY real estate buffoon, exiled to the world of reality television and catchphrases. He was a buffoon willing to say anything for attention, and we revelled in his buffoonery.

He told Robert Pattinson that Kristen Stewart wasn't good enough for him, and we laughed.

He made fun of rival TV show ratings, and we laughed. He said Barack Obama wasn't born in the United States, and we laughed.

Except we didn't all laugh. The idiot, the carnival barker, the fool was tapping into America's own original sin: racism. And he was building an audience. A large audience. And like any self-obsessed paranoid sociopath he reveled in the attention. And he kept doing the thing that got him the attention. And the more attention it got him, the more he did it. And the meaner he got the more attention we gave him.

Soon tweets about TV appearances and celebrities were replaced by tweets about birtherism, the Central Park Five, tearing down women, Muslims, and other assorted targets of hate. And his audience grew.

And at some point, and I don't know exactly when or how or who—even scarier I don't know if the people involved know when or how or who—Twitter made the decision to ride the hate wave. With their investors demanding growth, and their leadership blind to the bomb they were sitting on, Twitter decided that the audience Trump was bringing them was more important than upholding their core principles, their ethics, and their own terms of service.

And that, whenever that day might have been, is the day Twitter died.

...

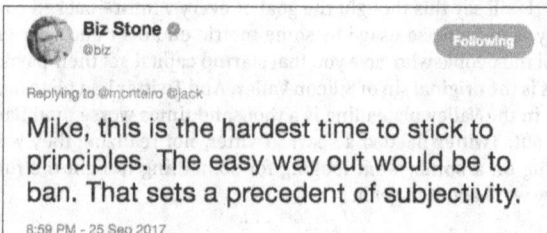

Twitter would have you believe that it's a beacon of free speech. Biz Stone[5] would have you believe that inaction is principle. I would ask you to consider the voices that have been silenced. The voices that have disappeared from Twitter because of the hatred and the abuse. Those voices aren't free. Those voices have been caged. Twitter has become an engine for further marginalization of the marginalized. A pretty hate machine.

Biz Stone would also have you believe that Twitter is being objective in its principled stance. To which I'd ask how objective it is that it constantly moves the goal posts of permissibility for its cash cow of hate. Trump's tweets are the methane that powers the pretty hate machine. But they're also the fuel for the bomb Twitter doesn't yet, even now, realize it is sitting on. There's a hell of a difference between giving Robert Pattinson dating advice and threatening a nuclear power with war.

...

In an alternate universe, that tweet was immediately followed up by a press release from Twitter CEO Jack Dorsey:

---

5 No one's really sure what Biz does at Twitter, other than he's able to smile and shake hands at the same time, which, to be fair, makes him stand out within that leadership team. But if you're ever looking for an example of a white dude who keeps succeeding without any apparently good reason, Biz is your man.

> "Twitter wholeheartedly condemns the recent tweets by Donald Trump toward North Korea. Twitter is a platform for global voices to come together. And though we are fierce advocates of free speech, we cannot stand idly by while our platform is used for harassment of another human or a sovereign nation. We expect all our users to abide by the terms of service they agreed to when they signed up for accounts. For this reason, we are putting a 24 hour ban on the @realDonaldTrump account. Once the offending tweets are deleted the account will be reactivated in 24 hours. Should the same offense recur, we reserve the right to institute another temporary ban or ban the account completely. These rules apply to all users. Cheers, Jack Dorsey, Twitter CEO."

That's what principles look like. Didn't happen.

...

Twitter today is a cesspool of hate. A plague of frogs. Ten years ago, a group of white dudes baked the DNA of the platform without thought to harassment or abuse. They built the platform with the best of intentions. I still believe this. But they were ignorant to their own blind spots. As we all are. This is the value of diverse teams by the way. When you're building a tool with a global reach (and who isn't these days) your team needs to look like the world it's trying to reach. And ten years later, the abuse has proven too much to fix.

I've known plenty of people who've worked at Twitter over the years. Most have left by now. Usually out of frustration. And their stories aren't mine to tell, so I won't. But I'll tell you this: a lot of those people have tried, honestly tried to deal with the abuse on the platform. But when leadership doesn't want something fixed, it's close to impossible to fix it. And when leadership doesn't see something as a problem, it's not getting fixed at all.

And I'm sure that in the next few days Jack Dorsey will come out and make a pledge about how Twitter needs to be more transparent.[6] He's very good at that. But when companies tell you they need to be more transparent, it's generally because they've been caught being transparent. You accidentally saw behind the curtain. Twitter is behaving exactly as it's been designed to behave. Twitter, at this moment, is the sum of the choices it has made. Even when the coop is covered in chickenshit, the chickens will come home to roost.

---

6 *In 2017, Buzzfeed published a collection of Jack Dorsey's tweets where he promised to be more transparent. It was a rather long article.*

Twitter never saw Donald Trump as a problem, because they saw him as the solution. As Upton Sinclair so eloquently put it:

"It is difficult to get a man to understand something when his salary depends upon his not understanding it."

When I started writing this essay, I was wondering whether to close my Twitter account. To be honest with you, all the Twitter shit was bringing me very fucking close to a mental health red flag, which is a thing I need to look out for. I still haven't decided whether or not to close it.[7] But I have decided that in writing all this shit down and having said my peace, I can step back a little bit, at least for a little while.

This isn't about Twitter anymore. This is about something bigger. When Donald Trump tweets us into war, the bombs don't fall inside Twitter. When Donald Trump tweets us out of the social contract, citizens who've never used the service are left to suffer. What happens when the thing that might save you is also the thing that might destroy the world? What do you do? Where does your responsibility lie?

Twitter set out to change the world. It did.

---

7   *I did for a while. And then, like an addict, I came right back.*

HOW TO EXPLAIN TO YOUR CHILDREN
THAT YOU WORK AT FACEBOOK

(Originally published in Modus on February 7, 2020)

On March 14, 2018, millions of students across the United States, across all grades, walked out of their classrooms for the March For Our Lives. This was specifically in response to the Parkland High School massacre, where 17 kids were murdered, but also generally about America's inability to address gun control. (In 2019, 776 teens and 209 children were murdered by guns in America.) Kids were marching to solve a problem adults in America had failed to address.

On September 20, 2019, three days before the United Nations Climate Summit, roughly six million people, in over 4,500 locations in 150 countries, participated in the Global Climate Strike. This was the third global strike of the school Strike for Climate movement. It was organized by children and spearheaded by Greta Thunberg, who was 16 at the time.

This morning I woke up to a tweet about grad students at California College of Arts coming up with their own code of ethics for design. The first point in their code is "If you can't make it better, don't make it worse." The last point is "Design with your grandchildren in mind." Everything in between those two points is just as great. And while grad students can hardly be described as children, many haven't yet entered the workplace, so I'll include them in the wave.

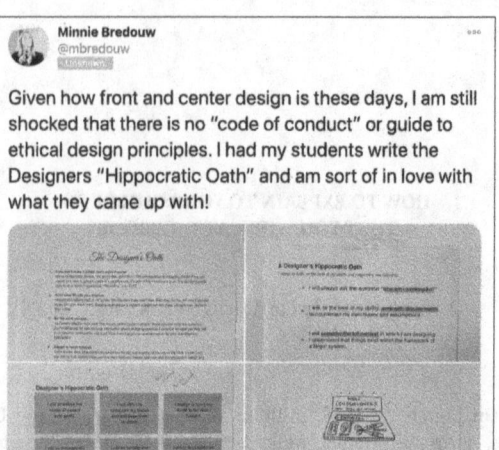

Minnie Bredouw @mbredouw

Given how front and center design is these days, I am still shocked that there is no "code of conduct" or guide to ethical design principles. I had my students write the Designers "Hippocratic Oath" and am sort of in love with what they came up with!

6:58 PM · Feb 3, 2020 · Twitter for iPhone

What I'm getting at here is pretty simple: One, the kids are all right. Two, the kids are angry. And they have every right to be. Not only are we handing them a broken world, but there's a very solid chance that we might not be handing them a world at all. And we've destroyed the world with the full knowledge that our actions were destroying the world. We did it anyway.

Going back to Greta Thunberg for a second—the amount of hatred being hurled at a 16-year-old by world leaders and captains of industry (including business leaders like Jack Dorsey and Mark Zuckerberg, who not only allow their platforms to be used to harass a 16-year-old but also profit from it) is not only shameful but also incredibly telling: we have no answer for what she's saying. She's right. And since we can't argue with the truth, we belittle the messenger.

Here's the thing though: While Greta Thunberg is the most well-known child in the climate crisis movement and is deserving of all the praise she is getting, she is not one of a kind. She is just the most visible part of the wave. She's the child the media has focused on, and she's risen to the occasion, for which she should be lauded. But she's hardly the only one who's angry at our actions. And she's hardly the one you should be the most worried about.

...

The Tech Bro Angers

Right before last year's Thanksgiving holidays, a time when Americans traditionally gather with their loved ones and their racist uncles, Facebook gave their employees a gift. It was called the Liam Bot. The purpose of the Liam Bot was to help Facebook employees navigate difficult questions from their family members. Facebook, which we used to think of as a place to share photos of our kids with the grandparents, has become the third rail of family conversations. At least for families of Facebook employees. (If that weren't true, they wouldn't have sunk resources into the app, Chad.)

Liam Bot is a chatbot Facebook employees could use to help them answer sticky questions like "Why does your company allow politicians to lie on the platform?" Which, to be fair, is such a difficult question to answer that Facebook's CEO doesn't actually have an answer for it.[1]

But where Liam Bot helps you defend attacks from the adults in the room, it does nothing to help you where your defenses are weakest: your children. Some of you may already have children old enough to ask you questions. Some of you may have toddlers who will one day ask you these questions, and some of you may be thinking of having children some day, and worried that they'll ask you these questions. They will.

At some point you will have to explain to your children that you work, or worked, at Facebook. For example, Chamath Palihapitiya, who was once Facebook's vice president for user growth, has become a very vocal critic of social media: "The short-term, dopamine-driven feedback loops we've created are destroying how society works."

I agree with him, by the way. And he should know—he helped build those things at Facebook. He's also said that his children "aren't allowed to use that shit." I imagine at some point Mr. Palihapitiya's children may ask him why he chose to work there, if they haven't already. They may even ask a follow-up, such as why was he okay with building things that addicted their friends.

In the interest of helping some Facebook parents answer some of the questions their children might ask, I've come up with some possible topics which might arise in the next few years. Liam Bot developers can feel free to add these to the next update. You're welcome.

---

1 *This is a reference to the many times Zuckerberg has short-circuited in front of Congressional Committees. His eyes glaze over, and he starts sweating that milky shit Ridley Scott robots are made with. For someone who so clearly believes he's speaking to people beneath him, he sure has a hard time answering their questions. He's not smart.*

WHEN YOUR CHILDREN ASK WHY YOU'RE HELPING TO ELECT A CLIMATE CRISIS DENIER

Yeah, this one's gonna hurt. First off, let's deal with the facts. The climate crisis is real (I'm not gonna even bother including a support footnote for that because I don't want to insult either of us). Trump is a climate crisis denier. He's pulled out of the Paris Accords, cut environmental protections multiple times, and has tweeted as much. And according to your own people, you helped get him elected. You're not going to be able to answer this question by disputing the facts, because your kid will be right.

If there's another way to tackle this question, I'm not sure what it is. So let me ask you: Why are you helping to elect a climate crisis denier? Your kids are gonna need a planet, and all their stuff is on this one. All those hopes and dreams about your kid learning how to ride a bike, graduating, getting married, giving you grandchildren to spoil, etc.—they all have one thing in common: they all happen on this planet. And yet you sell your labor to a company that caters to a psychopath who's doing everything possible to destroy that planet. Perhaps you don't even like working at Facebook, and you're doing it for the money, to ensure your kid has a good future. Sure, who doesn't want that for their kid? But a fat college fund is useless when the planet has been destroyed by natural disasters and civil wars for resources.

There must be times when this keeps you up at night.

WHEN YOUR CHILDREN ASK WHY YOU'RE HELPING TO ELECT A RACIST, SEXIST, MISOGYNISTIC XENOPHOBE

I mean, this isn't so different from the point above, and we certainly don't need to go over the facts, do we? I'm not going to spend my limited word count convincing you the man is who he shows you to be on a daily basis. Instead, let's do a little exercise. The next time you're at a kid's birthday party, or at a playground with your kid, or anywhere kids congregate, take a look at all the kids. Make a tally of how many of those kids Donald Trump, the man you're actively using your labor to re-elect, believes don't belong on that playground or at that birthday party.

Then take a look at your own kid. See how happy your kid is to be playing with those kids? Someday your kid is gonna ask why you work at a company that's helping elect someone who doesn't want their friends around. I surely hope you have an answer better than, "Well, we're protecting free speech." Especially if your kid comes back with "Well, what about Manny's free speech?" Brutal.

I mean, that Cybertruck you bought to own all the other parents at soccer practice isn't gonna be so exciting with all those empty seats.

WHEN YOUR CHILDREN ASK WHY IT'S OKAY TO LIE

Of all the important lessons a parent has to teach a child, honesty might be the most important. We'll all make mistakes in life; that's part of being human. We teach our children that you own your mistakes, you admit to what you did, and you always tell the truth. It's one of the most important cornerstones of being a decent person.

At some point your child will test that boundary. It's their job to test boundaries. They'll tell some small fib. It'll be obvious that they're telling a fib. Their goal isn't to get away with it, it's to see how you react. And you will do your job, as a good parent, and say something like "We don't lie in this house."

And then your kid will say, "But you let people lie at work. Why is that okay?"

Your kid will be right. You do let people lie at work. Mark Zuckerberg has stated over and over that he's okay with publishing political lies. He's even going on tour about it, he's so excited.

You may even tell your kid, "Well, Daddy doesn't work on the team that does that." Congratulations, you've just taught your child about money laundering.

...

Look, I know how much all of you love your kids, or will love your future kids. And I get that you want to ensure they have a good future. Every parent wants that. But when you're working at a place that's designed to destroy the future, doing that company's bidding cannot, by definition, ensure the future. Love your kids. Love them enough to stand up to those who'd take their future away. If you made it this far down the article you're probably pissed off at me. That's okay. But at least admit that you're pissed off at me because I'm reminding you of things you don't want to be reminded of. I understand that I may not have the right to remind you of those things. But your children do. And someday they will.

Ensure their future. For real.

THE PEOPLE VS DONALD TRUMP VS TWITTER

(Originally published on Medium on September 28, 2017)

We need to get Donald Trump off Twitter before he gets us all killed.

On September 23, 2017 he declared war on North Korea.

Now, US Presidents declaring war is nothing new. It's a thing they do. (Although Congress does like to be somewhat involved.) But from what I remember of history, this is a power that Presidents used to take very seriously. Both Roosevelt and Wilson famously wavered back and forth on whether to enter the World Wars. And, of course, Lincoln's nightmares about (rightfully) taking the US down a path that was guaranteed to start a Civil War is the stuff of legend. Normally, the power to take a nation to war is not to be taken lightly.

Our current president did it in a tweet:

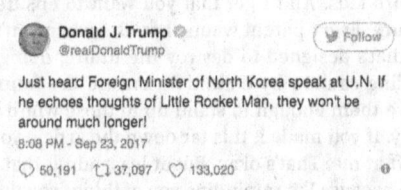

Donald J. Trump
@realDonaldTrump

Just heard Foreign Minister of North Korea speak at U.N. If he echoes thoughts of Little Rocket Man, they won't be around much longer!

8:08 PM - Sep 23, 2017

♡ 50,191   ⟲ 37,097   ♡ 133,020

Now you can argue whether that is, in fact, a declaration of war (some of you will pedantically and annoyingly argue that technically we are still at war with North Korea[1]), but in the end, what matters is whether the other nation sees it that way. In this case, that other nation is North Korea. And they, in fact, saw it as a clear declaration of war. Ultimately, the people who launch the bombs

---

[1] *This parenthetical was written specifically to troll my friend Ross Floate.*

have final say. And ultimately, the people those bombs are aimed at pay the final price.

Those are the people I worry about. And selfishly speaking, I worry that me and mine might be included in that group. But feel free to replace me and mine with you and yours because we're all fucked. We have an unstable president. And the first thing you do with unstable people is you take away their tools for hurting themselves and others.

Donald Trump has been violating Twitter's "rules" for years. Calling out individuals, entire ethnic groups, dogwhistling his violent white supremacist base, taking on a Gold Star family, a US judge of Mexican heritage, retweeting a gif of Hillary Clinton being attacked, going after journalists. This is hardly acceptable behavior for a regular human being, much less a US President. And Twitter has, rightfully albeit slowly, banned other users for similar behavior.

So, why hasn't Donald Trump been kicked off for his violent threats? He agreed to the same terms of service we all did when he signed up.

On September 25, Twitter co-founder Biz Stone finally addressed the issue in a short tepidly-written piece on Medium. Here's the interesting and scary as shit part:

> *"This has long been internal policy and we'll soon update our public-facing rules to reflect it."*

It's almost a throwaway line. But it holds a very important key to Twitter's thinking. In short: we will move the goalposts accordingly. Twitter is updating their policy to explicitly avoid banning Trump, while telling you that Donald Trump is held to the same rules as all other users.

After reading Biz' piece on Medium I told him it sounded hollow. Actually, to be completely transparent, I told him it was "100% bullshit". His reply:

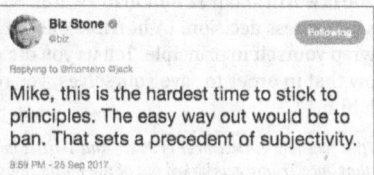

Changing the rules to fit the situation is the exact opposite of principles, and the exact meaning of subjectivity. It's cowardly. And it's opportunistic. Biz also argues that Trump's tweets are newsworthy. And I don't disagree. It's newsworthy to see if a president starts a war on your platform. In fact, it's a media coup. Take that, Facebook! Alas, Twitter isn't a media company, if they were they'd know that the function of a media company isn't to create news. The "newsworthiness" value is a newspaper media value. It has meaning in that setting.

According to my friend Ross Floate, who unlike me, actually has a journalism degree, that meaning is derived from three values—significance, timeliness, and proximity. "Newsworthiness" doesn't mean "this is used to break news." Using the value as they describe it makes all hate speech protected—if I threaten to murder you and then follow through, that's certainly newsworthy. Proper "newsworthiness" is not something that an algorithm does on the fly. It's a human value judgement. Also, the "newsworthy" argument isn't in their terms of service either. According to Biz Stone's Medium piece, they'll be adding it soon. Again, they're updating their policy to explicitly avoid having to ban Trump.

I'm assuming that one of the principles Biz believes he is referring to is free speech. I'm a big fan of free speech as well. But free speech isn't free of repercussions. Also, the First Amendment, which is very short and to the point, states that the government can't restrict your speech. It doesn't apply to Twitter, which is a private service. You know how bartenders can kick you out of a bar for being a belligerent dick? They have a right to do that. You don't have the right to be a belligerent dick in their bar. Twitter has a right to kick you out of their private service for acting like a belligerent dick too. Donald Trump is the asshole in the corner screaming at everyone and grabbing every woman's ass and telling the bartender not to serve the darkies.

Twitter believes it's protecting free speech, but it's really protecting that belligerent dick at the bar, at the expense of the other patrons. That's not a principled stance. That's a cowardly act.

I get that the Trump Twitter economy is strong. According to Fortune, Trump is worth about $2 billion to Twitter. So say that, Biz. Tell us this is a business decision. (A horrible one, to be sure.) But please don't wrap yourself in principle. Tell us you die without him.

Just know that in order to save yourselves, you're sentencing us to die with him.[2]

---

2 *I'm writing this footnote in December of 2020, and Twitter has recently announced that once Trump gets kicked out of the White House on January 20, 2021 (fingers so fucking crossed) they'll "be able to" hold him to the same rules as all other users. So... the rules that they put in place, and could've changed at any time, were keeping them from doing the right thing. Once they've milked Trump for all they could, they'd now appreciate a medal for tossing out the husk. Fuck these people.*

FACEBOOK ISN'T A COMMUNITY, IT'S A COMPANY TOWN

(Originally published on Medium on November 25, 2018)

A few years ago, I gave a talk on the Facebook campus. Campus is the most important word in that sentence. Corporations don't have offices anymore. They have campuses. And that's not an accident. The last time I was part of a campus, I was in college. I have fond memories of that campus. I grew up there. I met my friends there. I ate there. I saw bands there. I fell in love there. I had my heart broken there. I worked there. I shopped at the campus store. I even managed to attend a few classes there. And for the first year of my college life, I lived there.

For the length of my college stay that campus was the center of my community. And for people who attend college, this is the first community we experience outside the community in which we grew up. It's our first community as supposed adults, where we have a higher degree of agency.

"We're hard wired for community. We've been living in communities forever in terms of needing the numbers to go hunt or to go collect berries." says Dan Sinker. Dan has been building communities since he started Punk Planet, a magazine I probably read on that college campus, a practice he continued with Open News, a community of news nerds.

As Dan suggests, communities keep us safe. They protect us from elements outside the community. They allow us to share resources, and they provide us with an identity. Which is why so many people return to those college campuses throughout their lives hoping to rekindle that feeling of belonging.

Throughout our lives we continue looking for communities to which to belong. For some of us it's our neighborhood, or people with whom we share a common interest. Such as all the neighbors at the dog park whose names I don't know even as I can give you a full run-down of their dogs' dietary restrictions. A community can be a group of people with whom we share a spiritual connection, such as a church, synagogue, mosque, or a Butthole Surfers show. And sometimes community is thrust upon us. People marching for a cause are a temporary community. We have a sense of belonging together. In all these cases, these are our people. We look out for each other, and we keep each other safe.

As I walked through the Facebook campus, it reignited some of those same feelings from my college campus. There were shops, giant lawns, a climbing wall (my state school did not have a climbing wall), there were people doing things that looked more like hanging out than doing work. And people were wearing clothes with the name of the company on it, just like on college campuses. (Picture John Belushi in the infamous COLLEGE sweatshirt from Animal House, except it says COMPANY.) And just as I went to school to supposedly go to class, these people were here to supposedly do work. Yet the setting provided for so much more than that. It was obviously a community.

This is by design. Facebook, and companies like Facebook, want you to feel like you're not just at work, they want to be your de facto community. They'll provide you with everything you need. Not just a job, but also food, clothing, services to wash that clothing, social events, haircuts, they'll bring in bands to play at events, gyms, health care, and even on-campus mental health services (which raises so many red flags to me that it's beyond anything I can joke about).

Modern tech campuses don't just rival college campuses, they obliterate them in scope, activities, and money. Losing a job doesn't just mean losing a paycheck, it means being ostracized from your community. And in at least Facebook's case, losing access to your therapist! They're company towns.

There are communities which emerge, often without leaders (Occupy comes to mind), and then there are communities which are purposely designed to be communities, such as college campuses and these new tech company towns. When something is designed, we need to look at the motives of the designer. Tech campuses are designed to, first of all, lure you in. (Not much different than a college campus, to be honest.) When you interview, you visit the campus. It plays a role in which job you choose.

First they lure you in, and then they're designed to keep you there. (And here we diverge from college campuses, which are happy to kick you out after four years to groom to a new crop of future alumni dues targets.) They cater to your needs and whims. They provide sustenance. They provide necessary services.

Tech campuses are designed to lure you, to keep you there,

and most insidiously, they're designed to inspire loyalty. Especially when the community is under attack. They may appear to be designed for the benefit of the worker, but the feelings of loyalty the community is designed to engender benefit the company much, much more.

After the Facebook opposition research scandal[1] broke last week I reached out to a few Facebook employees. (They'll remain anonymous for obvious reasons.) And while their initial reactions went from pretending this wasn't happening, to being outraged and wanting to leave the company, within days they'd circled the wagons. Their mindset turned to something more like "we'll be fine. We'll get through this." They were rallying in support of what they saw as their community. And by the insidious design of that community, they ended up protecting the corporation that designed it.

There's a dark side of communities as well. And that's insulation. When the reason for community changes from 'keeping those inside safe' to 'keeping those outside, out" we lose perspective. We stagnate, and we stop introducing new ideas. (Not to mention crippling the genetic pool.) We circle the wagons.

According to the Intelligencer's Brian Feldman, Mark Zuckerberg gathered his staff together shortly after the Facebook opposition scandal broke and described the story as "bullshit." (Narrator: It's not. They ended up admitting it on the Wednesday evening before Thanksgiving.)

Zuckerberg is responsible for at least two communities: The community of people who use his service, and the community that builds his service. He betrayed the former a long time ago, and I can't be absolutely certain of exactly when he betrayed the latter, but that particular moment is definitely on the short list.

Where most of the world saw a story about corporate malfeasance and corruption, Facebook employees were told a story about their community being under attack and needing to protect itself. They circled the wagons. If your community was under attack, you'd circle the wagons too. And that's the problem. Their sense of community, as designed by the corporation, was stronger than their sense of loyalty to any community outside the company. (And

---

1 *So my editor left a note here: "exactly which Facebook scandal was this?" And for the life of me I couldn't even remember. There've been a lot. Which is kinda the point. When you ask someone about "the Obama scandal" they look at you all funny and say "What, the tan suit? You're tripping." But if you mention "the Trump scandal" they get a very tired look on their face. They may even start crying. They most definitely look for a place to sit, because they know this will take some time. Facebook is like that.*

the benefits of being in that community so immense!) It became easier to rally against outside forces, than to question the community which provides them with nearly everything. Not to mention their own complicity by continuing to work there.

Many of those same Facebook employees will end their day by getting on luxury buses and riding up highway 101 to San Francisco, a city where they sleep, eat brunch, and drink. But many of them don't view the city as their community. And that's the problem. The tech bubble, which eviscerated the rental market and led to thousands of evictions within the city, didn't just destroy multiple communities. It replaced them with non-communities. San Francisco, once a vibrant imperfect city with a million weird and wonderful communities, has become a bedroom community for Silicon Valley. A huge swath of people working here don't see themselves as part of any community in the city. Their community is elsewhere. And it was designed by their boss.

If designers, and other tech workers, want to have any chance at fixing the mess we've created, we need to reassess who we consider our community. The homeless people whose existence we condemn in our Medium think pieces because they dare to exist close to the homes we pay too much rent for? They are our community. And it would serve us all well to understand how we are failing them. The multitudes that get harassed and abused online by the very tools we build? They are our community. They deserve our allegiance. The corner bodega that's barely getting by because all their customers have been evicted? They deserve our business. The school system that's suffering because teachers can no longer afford to live where they teach? They deserve your tax money.

Now is it possible for real community to exist within the Disneyfied campuses created by corporations? I wouldn't be asking the theoretical question if the answer was no. Obviously, it can. Communities can emerge anywhere. They can emerge wherever people have common goals. Whether it's getting everyone on our block to rake their leaves in the fall, cheering our local team, or marching for a cause. We see this in the Google walkout just a few weeks ago. Google has many beautiful campuses across the world, and I guarantee you they weren't designed so that workers can organize. And yet they did. Those workers ended up creating a real community within the community the corporation designed for them.

The people who would sack you in a heartbeat to improve their quarterly earnings report are not your community and they don't deserve your allegiance. (And, by the way, when they fire you so they don't get yelled at during an investor call, they're showing you who their community is. It's not you.)

When companies design their workplaces to mimic the trappings of community—and many tech workers go directly from college campuses to corporate campuses, making the mental mapping that much easier—they're taking all those good feelings that you have about community and transferring it to their own ends. They're buying your loyalty with corporate haircuts, swordfish on Fridays, and a therapist within walking distance of your desk (that your manager can see you walking in and out of).

One of the reasons humans band together in larger communities is to protect each other from something larger than ourselves. Our power derives from our collective power. We can't design things for the common good if the community we're representing is our boss. When we look out at a team with twelve people in it, we need to know that team is representing the best interests of twelve different communities. We need to know that team is doing something those communities need. We need to know that team is making sure that tool isn't going to have adverse effects on those communities. Will that slow us down? You bet. And that's a feature, not a bug. We've seen where moving fast and breaking things has gotten us.

We need to get our shit together. Our real communities need us.

MERRY LAST CHRISTMAS, JACK DORSEY

(Originally published on Medium on December 20, 2017)

On Wednesday, November 29, 2017, Donald Trump, the president of the United States, woke up and started his day by retweeting three anti-muslim videos from a British fascist organization. To say he did this because he agreed with the viewpoints expressed in the videos isn't controversial. It may be somewhat controversial to say this makes him a fascist sympathizer. But only if you haven't been paying attention. Donald Trump is a fascist sympathizer.

This happened on Twitter. Again, that's not a shock. That's where Donald Trump engages his fan base. (And leaders of foreign nations like North Korea.) It's where he's his most unhinged.

Twitter reacted to the retweets of November 29 and their subsequent backlash by tripping over its own dick. Again, not a shock. For a company that traffics in outrage, Twitter appears to have a very hard time figuring out what to do when the outrage it covets shows up at its door.

First, a Twitter representative tweeted out that the videos, which were extremely violent in nature, didn't violate their rules because "...to help ensure people have an opportunity to see every side of an issue, there may be the rare occasion when we allow controversial content."

Oof.

A day later, Jack Dorsey tweeted out "we mistakenly pointed to the wrong reason we didn't take action on the videos from earlier this week." He then claimed the videos didn't violate their current media policy. Without elaborating on what that media policy was. My guess? He has no idea what the current media policy is. He still doesn't. He never has. He then promised to roll out a new set of rules designed to curb the amount of hate groups, hate symbols, and harassment on the platform.

Those changes were rolled out on Monday, December 18.

Among the changes, swastikas are now banned from Twitter. That's a good move. I applaud it, and it's beyond time. However, the Confederate flag, a hate symbol that defines one race's desire to own another race, is still acceptable. Twitter's reason is that the Confederate flag is historical. But so is the swastika. This decision seems less based on principles, but more on a desire to not piss off a certain group.

Despite their sanctimonious appeal to "principles", Twitter appears to be making decisions based on who they're afraid to (or can't afford to) piss off and then backward engineering the rationale to make it palatable. That's not principled. That's cynical.

Narrator: It's not the hardest time to stick to principles.

The new policy also allows a thinly veiled loophole for Trump:

"This policy does not apply to military or government entities, and we will consider exceptions for groups that are currently engaging in (or have engaged in) peaceful resolution."

For all his promises about wanting to reduce the amount of hate and violence and harassment on the platform, Jack Dorsey has purposely built in a loophole for the biggest perpetrator of all those things. That's not principled. It's cowardly. And it's dangerous. We've got a sitting US President using a social network to provoke a war with a foreign power.

Jack Dorsey has promised changes to the platform before. Usually after an embarrassing episode where a celebrity gets harassed. Hey, if that's what it takes, fine. Unfortunately, the promises quickly turn into vaporware, or smoke and mirror offers of rounded avatars, more characters to harass people with, and threading. None of which solves Twitter's main problem, which is that it's a nuclear bomb powered by a cowardly unprincipled boy king working part-time.

It's time to judge Jack Dorsey by his actions and not his promises. And his actions are lacking.

A year ago, when I started tweeting with Jack about this shit, he agreed to meet with me in person. I give him a lot of credit for doing this. And we met. It was respectful. My mother taught me how to be a good guest in someone's home. And he was a good host.[1] We discussed all the fears I had about how Trump and the fascists (I refuse to say alt-right, that's a bullshit term) would use Twitter to harass and silence others. We discussed Twitter's role in the world stage.

And I admired his vision, but feared his approach. Jack, and

---

1 *Fun fact: he insisted on meeting at Square, you know—his other company, rather than at Twitter HQ.*

to an extent Twitter's pet porg Biz Stone[2], have always believed that absolute free speech is the answer. They're blind to the voices silenced by hate and intimidation. The voices that need to be protected. But anyone who's ever tended a garden knows that for the good stuff to grow, you have to deal with the bad stuff. You can't let the weeds choke the vegetables. You'll go hungry.

I walked into that meeting worried that Jack didn't have the leadership qualities to guide Twitter through the shitstorm to come. I walked out of it assured of that. I walked away from that meeting feeling like Jack was in over his head. Time has borne that out.

Few people on Earth have the experience to deal with a company of this scope. But I can't help thinking that someone from a marginalized group would at least have a better idea of the danger signals. I can't help but draw a comparison to George W. Bush whose lifelong dream was to be baseball commissioner. Instead, he was pushed to being an ineffectual president. Jack came very close to leaving Twitter in its infancy to become a dressmaker. I believe Jack would've been a good dressmaker. (This isn't a slight. My mother's a dressmaker.)

Twitter isn't a technology company. It's a human interaction company. I'm not sure they've ever understood that. And that's the generous assumption. More likely, and alarmingly, they understand that but don't care. And while a lot of this falls on Jack's head—after all his name's on the shingle—you also need to take a look at the people who put him in that position. The Twitter board has allowed this to happen. They are also complicit. Their job is to help the captain steer the ship. To offer guidance. One wonders if their guidance has been "fascists or not, they're driving engagement," because that seems to be Twitter's governing force. So it's fair to examine the governing body.

It's also fair to look at the overall system in place. Because Twitter isn't an isolated case. The ethical chickens of Silicon Valley have come home to roost this year. And in most situations what we're seeing is that Silicon Valley, which is very good at evaluating technology, sucks at evaluating human interaction. They're very good at pushing their companies toward how to do things, but seldom ask whether those things should or shouldn't be done. We're seeing the results of that now, as companies like Twitter and Facebook are coming to terms with what exactly it is they broke by moving fast.

---

2   Brutal.

The Tech Bro Angers

Silicon Valley's time is almost up. So Merry last Christmas to them.

> *"Historians have a word for Germans who joined the Nazi party, not because they hated Jews, but out of a hope for restored patriotism, or a sense of economic anxiety, or a hope to preserve their religious values, or dislike of their opponents, or raw political opportunism, or convenience, or ignorance, or greed. That word is Nazi. Nobody cares about their motives anymore."*
>
> — A.R. Moxon

Yesterday, Charlie Warzel of Buzzfeed published an excellent piece about Twitter's inability to interpret its own rules while dealing with another fascist, Milo Yiannopoulos back in 2016. (This is a common theme for Twitter.) You should read it. In it, he quotes a Jack Dorsey email where Jack is quoting Gandhi to his employees: "You must not lose faith in humanity. Humanity is an ocean; if a few drops of the ocean are dirty, the ocean does not become dirty."

Here, Jack is right. Or rather, Gandhi is right. We mustn't lose faith in humanity.

But Twitter isn't humanity. It's a venture-backed tool that profits off human connections without really understanding how they work. It's not an ocean. The more apt metaphor would be a bloodstream. And while a single turd doesn't spoil an ocean, a single cancer cell in the bloodstream will spread and corrupt every healthy cell it encounters. For too long Jack has allowed the cancer of fascism to spread through Twitter's bloodstream.

For those arguing that it was Cornell West that drove Coates off Twitter, I'd submit it was Richard Spencer's tweet that drove him off, not Cornell West.

As of this writing, despite Twitter's new anti-harassment rules, Richard Spencer, Mike Cernovich, Joey Gibson, and David Duke are still on the platform. All known fascists. Ta-Nehisi Coates is not. Driven off by Richard Spencer's harassment. These are the voices we are losing because Jack Dorsey has chosen not to act. And it is probably too late. That's not hyperbolic. Jack has given Donald Trump the fuse to light a nuclear war. Jack has given Trump's minions the fuse to silence the voices America most needs at this moment. And Silicon Valley has allowed this to happen because their libertarian allegiance doesn't extend beyond their own pricks.

Merry last Christmas, Jack. To paraphrase James Baldwin, you have not earned your death. Yet you've mortgaged all of ours.

Your inability to make a courageous decision has left us with precious few to make ourselves. Whether it's fair to blame you for the last Christmas we all spend on Earth together or not, Jack, I firmly believe that you had an opportunity to stop it. Sadly, you didn't have the fortitude to do so. You've failed. Please know you've fucked the world.

Have yourself a very Merry Christmas, Jack.

THE 2020[1] GUIDE FOR WHITE MEN IN TECH

(Originally published in Modus on January 30, 2020)

Welcome back, gentlemen. Last year's guide to white men in tech was one of the most popular articles on this platform, so I'm excited about this year's. We assume the popularity was for several reasons: the title had a number in it, it was addressed to white men, it was written by a white man, and it was about tech. That's a Medium grand slam!

(I'm kidding. There was no guide last year. Roll with it.)

Look, I'm walking a tightrope here, kids. First of all, let's acknowledge, as white men, the privilege we have in society in general and in tech specifically. Then let's acknowledge that you get that privilege whether you want it or not. You can be super conscious of your privileged place in the world, you can disavow it, you can attempt to cast it off, you can rail against it. And I encourage you to do all of those things and more.

But at the end of the day, you're still going to get it. You're still going to be offered more and better jobs and get paid more for them, the bank won't call the cops on you if you go to deposit a check, the passport office won't give you side eye, and your co-workers won't confuse you for the building's custodial staff. Finally, I should acknowledge that I couldn't say half this shit and still expect to have a regular column here if it weren't for that privilege.

Privilege is unshakeable. Until the world is fundamentally changed so that all people are viewed as equal—which is our goal—you have two choices: You can either float along thinking you hit a triple when you really just happened to be born on third base, or you can use that privilege to help others who have to work much harder than you to achieve even half your success. As the recipi-

---

[1] *In my defense, I wrote this before we knew what a massively depressing garbage fire 2020 was going to be. I thought I'd start the year off with a little light-hearted satire. Yaaaay.*

ent of unfair advantage, it's on you to tip the scales back to where justice would have them. This will not make you a hero, but it will bring you closer to being a decent human being—which is a pretty wonderful thing to be, I've heard.

IT'S STILL A GREAT TIME TO BE A WHITE MAN

Your successful white father had it pretty good. Maybe he worked in finance, wore an Armani suit with big shoulders and swank suspenders. Carried a big brick of a mobile phone. A real Gordon Gekko vibe. He'd wheel and deal in the morning, do a little cocaine at lunch, go to some afternoon meetings, then hit the clubs. Sometimes he "worked so late" he stayed in the apartment he kept in the city.

Your successful white grandfather would take the train in from the 'burbs, do some office work, have a two-martini lunch, do a little more work, have his secretary buy his wife an anniversary present, as well as a little something for herself, and then head back to the 'burbs where the missus would be waiting with dinner on the table and a glass of scotch. Your successful white great-grandfather might have owned a top hat. Your successful white great-great-grandfather might have owned your coworker's great-great-grandfather.

Successful you has to work longer hours. You're saddled with a lot of student loans, and you're probably having dinner at work, or maybe picking it up on the way home because your wife has a job too. So, on the surface, it might look like things are on the decline even for the successful white man. On the surface. Until you take a look at society as a whole, and see that successful people of other races and genders are going through all of those same things while getting paid a lot less than you are. Plus you can get things like bank loans and VC funding and not get killed by cops during routine traffic stops.

The world may be slowly edging toward justice, on a generational level, but the last three years of US leadership is certainly pushing it backward. You're still living the high life, though. As the saying goes, "when you're accustomed to privilege, equality feels like oppression." Other people aren't subtracting from what you're entitled to. They're entitled to as much as you are.

I know you hate to wait as much as you love a microbrew IPA, so here are this year's bullet points:

STOP BEING SO EMOTIONAL ABOUT EVERYTHING

Despite your perception that things are now harder for you (they're not; we just went over that) you are going to need to hold your shit together if you want to be taken seriously. I understand how emotional white men can be, because I am one of you. I see you, Todd!

Sometimes people are wrong on the internet. Sometimes people are wrong at work. Sometimes the cafeteria is out of chicken molé enchiladas by the time you get there. Sometimes the company offsite can't be at a strip club because Karen made a fuss last time. Sometimes they remake your favorite childhood movies and they put women in them. I realize how traumatizing all of this can be, but my dudes—we freak out about this shit. We freak out when everything doesn't go our way because everything went our way for soooo long. And honestly, it still goes our way more than is fair. That's not a feature; that's a bug.

So, let's stop accusing others of being emotional at work when we're the ones spending our lunch hour rage-petitioning for the release of the "proper" Star Wars film edit.

SAVE YOUR ENERGY

Look, we all know that you knew how Mary-Anne was going to finish that sentence, but why exhaust yourself by finishing it for her? Let her tire herself out. Make her finish what she started. So, let's keep our pie-holes shut while Mary-Anne is talking. You know, to save our energy. After all, you have a century to ride this Friday. Oh, and maybe remind the other guys to save their energy too.

Also, promotions are exhausting. It's a lot more work, which cuts into "you" time, so the next time one comes up why not sit it out? Let Mary-Anne try to get it. Sucker. She'll be doing more work. And even if you end up reporting to her, you'll most likely still be making more money than she does.

SABOTAGE IS FUN!

Listen up, young Chads. Some of you are working at companies that are doing society a disservice. You may even feel bad about it. You've been telling your woke friends you've been "changing things from the inside" for years. Why not actually start doing it?

But rather than trying to organize your coworkers or collecting signatures for petitions, why not play to your particular white men strengths? Keep things from running smoothly. Introduce wedges into the cogs of the machine. Stick a potato in the proverbial tailpipe! Not only is sabotage ethical, it can also be a lot of fun. If your company is doing shady shit, you are in a great position to disrupt the flow of "progress."

If you're a worker, be late to meetings. Spill your coffee on the conference phone. Unplug random electrical plugs. Delete code for problematic projects. At all-hands meetings, ask long-winded questions that have little to do with anything. Address email to the wrong people. Don't document your work for colleagues who will need to use it or build on it. Take the batteries out of the remote for the presentation screen. Get needlessly emotional when there are no donuts left in the breakroom. Take excessive bathroom breaks and clog the toilet with an entire roll of paper. Do less work than the previous quarter and demand a raise. Take sick days as a deadline looms. Use your unlimited PTO until it becomes a problem, but come back before someone mentions it.

If you're a manager, call meetings that don't need to happen. Invite people who don't need to be there. Let meetings drift from one topic to another without resolution. Make sure there's always a remote caller and struggle with getting them set up. Tell callers you're using Zoom then wait for them in Hangouts. And make sure meetings run long enough so people are late to their next meeting—you get to disrupt two meetings! Promote incompetents and offer them no training. Assign your worst workers to the most problematic problems. Accidentally leak sensitive interoffice memos to members of the press. Lie to your boss about success metrics.

As white men, you can use your positions of privilege to sabotage the workplace and emerge relatively unscathed. You'll rarely be blamed, and if you're actually caught red-handed, it'll be assumed that you're under a large amount of stress and you'll probably be given extra vacation time to recharge, or promoted to a position where you can do less damage. If you're a company founder you can sabotage at will and get a $1.7 billion payout, as WeWork's Adam Neumann did!

EMPATHY IS FOR SUCKERS

You've probably sat through a couple of empathy workshops at this point. Probably something put together by the UX people and HR together. They're exhausting. Why should all this shit be on you? Why should you have to learn how to think like other people? Here's a loophole: you don't have to think like any of these other groups if you hire them to do the work for you. That's right! Why wonder how women and minorities might use something? Just let them build the damn thing. They can do all the hard work. They can do all the messy stuff.

You can just sit back, brother. Recline in that nice, expensive office chair. Put your feet up. See if McKinsey's hiring. And think about the bright future you and the other white guys built for everybody. Heckuva job, Dougie.

WHAT IF WE GET THROUGH THIS?

(Originally published in Dear Design Student on June 26, 2017)

Let me just put this here. Because I need a little bit of hope, and maybe you do too.

Because every time my phone vibrates, I expect to find out that we've launched a nuke, either into Southeast Asia, or the Middle East, or at a reporter's house. I expect that one branch of our government has declared war on another branch, or that members of the same branch are fat-shaming each other. I expect that an active gunman is mowing down children in a playground. I expect that US Senators are physically expelling Americans in wheelchairs from the Capitol. I expect these things because these things are actually happening. Things that would've freaked us out just five months ago are now barely worth taking note of. Because the next mess is hauling its way toward us just behind it.

And as mornings become more difficult to face, and the desire to find out how close we are to the bottom becomes stronger than the desire to keep fighting a daily barrage of disillusion in the rest of humankind, and SSRIs become the newest flavor of M&Ms, I wonder if maybe we need to be reminded of what it might look like if we made it through this shit show.

What if, despite all odds, we get through this?

It's said that in good times we crank out novels of dystopian futures to keep us humble. Perhaps the inverse also needs to be true. Perhaps we should clear out a little corner of the dumpster-on-fire, and pen some fan fiction to give us a smidge of hope.

What if, despite all odds, we get through this? What if, against all odds, we pulled out of this nosedive while there was still the chance, the opportunity, the responsibility to fix whatever's left.

What should we do?

LET'S FIX TECH

The service economy, where tech is putting the majority of its time and money, is built on income disparity. We create services for people to drive us around, to pick up our clothes, to sort our mail, and all of these services require the availability of two things: people who have and people who have not. The whole tech sector is based on exploitation of the lower class.[1]

How did we get here? We got here because once the internet turned into a business it was taken over by the people who run the business world. Rich white boys with business degrees who've relied on poor people to cook and clean for them their entire lives. You build what you know.

If we achieved income parity, the tech sector would collapse! You can't build an economy on the need for a class structure and then act surprised when it results in a class structure. Is this the kind of system you want to be supporting with the short amount of time you have on earth? Imagine if we put our time and energy into creating the infrastructure for an economy that actually leveled the playing field. An economy that served those who needed help the most first and foremost.

Isn't this socialism? Yes. Deal with it. But we've seen where the service economy leads. And we've seen where excluding everyone who isn't a rich white boy gets us. It's time to open the gates. The people solving the problems need to look like the world. They need to come from every corner of the world. And if you're investing your time and money in solving problems, it's time to start placing more value on how your work is affecting the world around you than on your own your own personal wealth.

We have more cognitive resources, computing power, and data available to us now than at any other time in human history. And we're using it to get poor people to do our laundry. Imagine taking all this intelligence, all this money, all this energy, and dedicating it to coming up with a plan that begins to make the world work better for everyone! And you're included in that everyone. Is it hippie dippy pollyanna? Sure, but it's my fan fiction.

---

1 Proposition 22, a ballot initiative on the California 2020 ballot was mostly bankrolled by Uber, Lyft, DoorDash, and Instacart. Its goal was to override a new state law that requires their ride-hailing and delivery drivers to be classified as employees rather than independent contractors. Uber spent $52 million to bankroll Prop 22, Lyft spent $49 million, Doordash spent $48 million, and Instacart spent $28 million. This was all money that could've been spent on worker benefits. But these companies' business models require exploited people in order to operate. Sadly, it passed. We suck.

LET'S FIX EDUCATION

A country that purposely undereducates its electorate in order to maintain its power structures loses the right to call itself a democracy. And an education system, which has historically underserved women and minorities, doesn't get to look out on a graduating engineering class that's 95% male and come to the conclusion that women aren't interested in engineering. You can't simultaneously cause a problem and use the results to justify believing the problem doesn't exist.

For decades we've been increasing the cost of higher education to levels which exclude lower-class kids (of all races and genders) from attending college. And those that do manage to attend, are saddled with so much debt that it dictates the degrees they're willing to get. (We've basically made humanities degrees extinct.) And the result has been a pipeline of self-perpetuating caucasian chicken nugget sludge. This is by design. We've created an undereducated electorate that we can bully with fear and anger. Democracies don't do that.

Imagine prioritizing education enough as a country that we shifted our tax dollars to guarantee that anyone who wanted that education could get it. For free. SF State is doing it. New York and Rhode Island as well. This has never been about lack of money, but about priorities.

LET'S DESTROY BRO CULTURE

When your product team looks like the photo below, you're telling everyone who doesn't look like this that they can't be on a product team. The prerequisite for being on this team isn't a skill set, or a level of education—it's being a dude.

We need women and minorities in positions of leadership. And we need to deal with the educational system so that we increase the diversity of people available to interview.

Then we need to deal with a bro culture that dismisses people on the basis of fit, which is nothing but a code word for misogyny and racism. Meritocracy? Don't even. That lie's been exposed too much recently. Not just by the good women we've kept down, but by the mediocrity that we've elevated to the top.

Imagine building teams that actually reflected the communities we're serving? Imagine having a trans person in the room when you propose a "real names" initiative? Imagine having someone who's been stalked in the room when you suggest adding places of employment to dating profiles. Heck, imagine having a woman in the room when you craft a healthcare bill! That's right, we don't get to make fun of Congress' shit when ours stinks just as bad.

It's time for designers to become grownups. We need a code of ethics and we need to get licensed.

LET'S FIX UX DESIGN CULTURE

The days of the wild west are over. When I started doing interaction design, I didn't have a degree. Because no one was teaching it. It was a brand new field. We figured out how to build the internet by building it. And it was great. But those days are over.

The stuff we're designing now is deeply enmeshed in the social fabric of our lives. We're designing things that put people in strangers' cars, that control devices in our homes, that administer medication. We design privacy settings that have deep repercussions in the ethical walls we've had to create for our messy, messy, wonderful lives.

When I started designing things on the internet, if you fucked up you got a broken link. Now if you fuck up, you mess up someone's life. And as much as I love doing my job, I'm beginning to think it's insane that I don't need a license to do this shit! I have no training in it!

And as our work gets more complicated, the ethical concerns about what we're doing get bigger and nastier.

(Look no further than all the shit we went through with Uber.) And I'm out there talking to a lot of designers. I'm genuinely scared at the amount of them who don't see ethics as a part of the craft. This needs to change. A doctor who gets busted working unethically loses their license. Can't practice medicine anymore. Same with lawyers. It's time for designers to become grownups. We need a code of ethics and we need to get licensed.

LET'S GET THERE

This is some of the stuff we need to fix. There's so, so, so much more. But this is the stuff that I think I can have a hand in trying to fix in my own field. And maybe, just maybe if we get through this current hell we can get a chance to tackle some of these.

...unless maybe, dealing with this shit now is how we can ensure we get through it. Maybe.

PART THREE

The Working Angers

(IF YOU WANT IT)

Gilly & Billy drawing by Adam Koford

> "Not everything that is faced can be changed, but nothing can be changed until it is faced."
>
> — James Baldwin

CAN DESIGN CHANGE THE WORLD?

(Originally published in Dear Design Student on October 30, 2015)

> *Q: I'll keep it simple: can design change the world?*

I'll answer just as simply: Yes. But...

But not for the reasons you think. And not the reasons you hear bandied about by the hopeful and the inspired and the lovingly crafted exuberant designers of today. Because I doubt very much that someone who can't change the boss's mind, and likely can't change a printer cartridge, can change the world.

Because I love you, I need to tell you something: you're not special. You have no unique properties. There is absolutely nothing about you that makes you different from anyone else. You are not a snowflake. And even if you are the most creative person you know, I guarantee there are ten million other people just... like... you.

This is good news.

I don't think design can change the world because it's special. I think design can change the world because it's not. Because I think anyone can change the world. And because the world isn't usually changed by special people. It's changed by ordinary people. Ordinary people who take it upon themselves to take a stand because they're trying to lead ordinary lives and something stupid gets in their way.

The world is changed by seamstresses who don't want to give up their seat when they're trying to get home after a hard day of work. The world is changed by electricians trying to earn a fair wage. The world is changed by girls who just want to go to school. The world is changed by low-level diplomats who just want to go home. The world is changed by cocky kids who find themselves punching above their pay grade.[1]

Let me tell you a story about a man who changed the world. At least my corner of it. His name was Bernard Harmon, and he was a public high school art teacher in Philadelphia. I was a freshman in high school when I entered his classroom. And like many of the kids in his class, I was having a pretty non-great childhood. I was a fat, pimpled, underachieving immigrant kid, dying to belong. And too angry to adjust. And looking around at that classroom, we were all pretty messed up kids. And all he had to do was just cruise us through the next four years of the Philadelphia public school system. Except he didn't. Instead he gave a shit. He became a de-facto dad for all of his students, especially the ones who needed him the most. He gave us goals, he made us work to achieve them, he pushed us, he gave us a moral center, and he made us grow up. And I guarantee you he was getting paid shit for all this because, like I said, he was a teacher in the Philadelphia public school system. And now the kids he taught are out there in the world doing their own thing, hopefully passing on the lessons we were taught.

Teachers can change the world. All around the world. Every day. In small ways, one kid at a time. And for little reward, other than something inside them that tells them it's the right thing to do.

> *"That's the problem with being middle-class. Anybody who really cares will abandon you for those who need it more."*
> — Mr. Bergstrom's parting remarks, "Lisa's Substitute"

Doctors can, and have, changed the world.
Journalists can, and have, changed the world.
And yes, designers can, and have, changed the world.

But it's not because we're especially capable of effecting change and improving the world around us. We're no different from anyone else. We're not special. We're ordinary. And we live by the same social contract. Yes, designers can change the world. But it's because we have the same responsibility as every other ordinary person.

So the real question isn't whether we can change the world, but how? What will you do? Can you afford not to? Greenland is melting. Children are being gunned down in schools. A great mass

---

[1] In order: Rosa Parks, Lech Walesa, Malala Yousafzai, Aristides de Sousa Mendes, and Muhammad Ali. Some of these are well-known, some are not. But that was kinda the point. This essay eventually turned into a talk, in which these folks portraits were displayed behind me as I went through the descriptions. It made a lot more sense that way.

of stateless people are dying as they pass through Europe. And there's probably a shelter in your own city which could use a volunteer to a teach an unhoused kid math.

So, no, you're not special. You're ordinary. And like everybody else, you need to opt in. We can't afford you not trying.

I wish you all ordinary lives.

YES, I WILL SHAME THE WORKERS

(Originally published in Modus on January 15, 2019)

"That's not a good look," came the reply.
"What's not a good look?" I asked.
"Shaming workers."

This was in response to a tweet I sent out a few weeks ago, pointing out some new horrible Facebook behavior. (There are many of these tweets to choose from in my feed. Pick one.) The tweet was addressed to Facebook workers and went something like "Facebook employees, this is where you work" with a link to an article about Facebook's newest horrible behavior. I've posted the same type of tweet to Twitter workers as well.

But still, I was taken aback by the accusation that I was shaming workers.

Was my intention to shame workers? My intention was to remind the workers at Facebook of the effect of their labor. What the effects of their labor cost society. As I've said many times, you are responsible for the work you put into the world. I firmly believe that, and I hope you do as well.

But also, it probably *was* my intention to shame those workers —if only a subconscious intention. So I will own it. Yes, I will shame workers. Because, Lord—some of these tech workers should be ashamed of what they are devoting their time to. And if they don't have the sense to feel that shame internally, I am happy to provide it for them. Being a Catholic alumni, I am more than qualified to do this.

Here's the thing about shaming someone, though: It's like watering a seed. The sun can shine all day long on dirt, you can water it all you want, but unless there's a seed in the dirt, nothing's gonna grow. If you tried to shame me about wearing my favorite baseball team's cap, it wouldn't work. I love my favorite baseball team. (It's the Phillies, BTW.) If you tried to shame me about watching too much TV, you might get a little traction. Because I probably do watch a little too much TV, but ultimately it would pass because I love watching TV. However, if you caught me smoking a cigarette and shamed me about it, that would totally work because it's a horrible habit and I shouldn't be doing it. (For the record, I don't. Not in a long time. But it's a good example most of us can agree on, so I went with it.)

Something tells me the employees of Facebook already feel shame. Just not enough to do anything about it yet. Every single employee at Facebook knows where the money for their paycheck comes from, and has for a long time. It's fair to assume they're okay with it.

Let's go back to our cigarette metaphor for a second. I smoked from college into my 40s. At no point did I think it wasn't an awful habit. At no point could I look at the evidence and decide that it was an okay thing to do. Luckily, I was alive during an era in which attitudes about smoking were shifting rapidly. It went from something most people did (I've smoked on planes!) to something that people tolerated others doing, to something we now ban from all places where people congregate. There was an undeniable realization that smoking harmed not just the smoker, but everyone around the smoker. Your actions harmed everyone around you. (This should feel familiar.)

The pressure to quit smoking came from society shaming me for smoking. And I am grateful for that. I am alive because of that. Would I have quit without the public shaming? Maybe. But I think we can all agree that society as a whole is healthier because of a coordinated campaign to publicly shame smokers and decrease the places they could smoke.

So, what's this got to do with being a tech worker? Easy. Addiction.

EVERYONE HAS A RIGHT TO EARN A LIVING

When a highly paid tech worker tells you they have the right to earn a living, there's a phrase missing: "...in the manner to which I've grown accustomed." You do have a right to earn a living, but so do the refugees and immigrants cataloged in the database tech workers at Palantir built for ICE. You don't have a right to earn a living by denying others their right to earn a living. And you don't have a right to earn a living that's one hundred times better than everyone else's living. The most important word in the phrase "Everyone has a right to earn a living" is everyone.

The upper echelon of tech workers earns a very good living, which isn't to say that there aren't tremendous problems with pay disparity along gender lines, and racial lines, and many other lines. There are. And those problems need to be fixed. But the minimum wage in San Francisco is $15.59/hr. Minimum wage in greater California is $12. Minimum wage in the United States is $7.25/hr. By those standards, you earn a very good living.

And yes, I realize that many of you are saddled with student debt and medical costs. That shit is real. But the solution isn't to do the corporation's work, it's to break the system in which a corporation can trade your debt for complicity.

And yes, those of you living in the Bay Area will say that your rents are high, and they are. They're definitely too high for anyone making minimum wage to afford. Tech workers complaining about the high cost of housing is not unlike a cancer cell complaining about a host body deteriorating.

You are not Bob Cratchett. You are not a Christmas ham away from your family dying of hunger. You have a right to earn a living, but not in the manner to which you've become accustomed. Your point is moot.

I don't have to take this bullshit from a privileged white guy!

Yes, I am a privileged white guy. I will own that. Doors open for me. I've never had to prove I belong somewhere. I've never been followed by a security guard when I walk into a store. I've never feared for my life when getting into a ride-share vehicle. I've never been afraid to leave my drink unattended at a bar to go to the bathroom. And I've never been hassled for being in a bathroom someone decided I didn't have a right to be in. I own all of that.

I get that many people reading this have climbed hurdles I never had to climb to get where they are today. And I respect that. But clearing those hurdles doesn't give anyone a pass to work on tools that abuse, harass, and spread lies.

Yes, I am a privileged white guy. And I use that privilege to yell at people. If you are going to check me on my privilege, check me for the many valid reasons available to you. Don't do it to mask your own shame about how you earn your money. And make sure you're not allowing Facebook to use you as a human shield. Be mindful that when you tell me to check my privilege you're not doing Facebook's PR for free. Corporations pit us against each other by design.

I'm not here for a seat at the table; I'm here to use the table for kindling.

WORKERS HAVE REAL OPTIONS

On December 31, 2019, 2,300 Google cafeteria workers unionized. Tired of being underpaid and overworked, the cafeteria workers organized themselves and reached out to a local chapter of Unite Here, which represents cafeteria workers, hotel workers, and other hospitality workers all over North America. They will now be able to bargain collectively, demand better working conditions collectively, and get paid what they deserve. Any of those things are monumental tasks when you act individually. But collectively their chances at success will improve.

THESE OPTIONS ARE AVAILABLE TO ALL WORKERS.

On January 7, 2020, the Communications Workers of America (CWA), one of America's biggest labor unions, launched a major initiative to unionize tech and gaming workers. Organizing is our best chance to do away with that seed of shame that's growing inside workers who realize their paychecks aren't earned in a manner they can feel good about.

Leadership will not change. They have no reason to. The current system is working great for them! WeWork founder Adam Neumann got a $1.7 billion payout for failing. That's privilege at work! And yes, the people involved in that transaction should be ashamed of themselves.

But they're not. And they never will be. They're not capable of it.

But you are. Or you wouldn't be getting mad at me right now.

That sense of dread that starts building on Sunday afternoon, knowing that tomorrow you'll have to go to work and build tools which harass people and lie to people and help cage people—that's shame. I'm not here to shame you. I don't need to. You're doing it to yourself. I'm here to tell you that you have more power and more strength than you believe you have.

THIS IS A HARD TIME TO BE A DESIGNER.

If you are afraid to fight, this might not be the right time for you to be a designer. If you are afraid to do the job ethically, this might not be the right time for you to be a designer. If you are afraid to speak truth to power, this might not be the right time for you to be a designer. If you are afraid to be judged by the impact of your work, this might not be the right time to be a designer. If you are afraid to stand up for the ones who need you the most, this might not be the right time to be a designer.

Find another way to earn your living. Tap out. Let there be no shame in that.

But if you are ready to do the right thing, look around you. There are others who feel the same way. Organize. You wanted to change the world? Here's your chance. You wanted to disrupt? Disrupt the system that's covering you in shame.

STOP COVERING YOUR ASS

(Originally published in Modus on September 5, 2019)

You keep talking about how designers have to take responsibility, but the real power within organizations is with engineers and product managers. All the shit is their fault, not ours!

Growing up a little Catholic boy, I was taught the most important of all the sacraments: the assignment of blame. And unlike the other sacraments, such as baptism and first communion, which had to be performed by a priest, the assignment of blame was the people's sacrament—it could be performed by anyone. In fact, it was encouraged. And no one was better at performing this ritual than my mother. My mother would enter a room while saying, "Who's to blame for..." and then quickly take in the scene before settling on a predicate, which she always did. There was always an opportunity to practice the holy sacrament, because there was always something that could be blamed on someone.

"I have a cold."

"It's your fault for not wearing a hat."

"I need new glasses."

"It's your fault for reading too much."

"I got beat up at school."

"It's your fault for having a smart mouth." (She was partially right about this one.)

Needless to say, growing up in a culture of blame will fuck you up, especially when blame comes with repercussions, as it often did. (I won't go into them here.) But you quickly realize the path to survival lies in not being blamed for things. So, what to do when the assignment of blame was non-negotiable? Luckily, I had brothers. And as any herd animal knows, when being chased by a predator you don't need to be the fastest in the herd, you just can't be the slowest. I got really good at deflecting blame for things onto my brothers. As did they. So growing up, we learned the key to survival was to make sure there was someone else who could be blamed. Our hands didn't have to be clean, they just couldn't be the dirtiest.

From what I've seen in the workplace, most of us grew up in some form of this dynamic, Catholic or not. (By the way, I'm an alumni now. But I still get to give them shit, which I enjoy.) When something goes wrong, our first instinct is to figure out who we can throw in front of the person looking to assign blame, be it engineers, product managers, a client, or someone in leadership. How often have you said or heard some version of "If/when this goes wrong it won't be my/our fault"?

Too often in the workplace our goal isn't to do good work, but to cover our ass. We even have an acronym for it: CYA. Cover your ass. Protect yourself from blame, possible criticism, repercussions, penalties, getting fired, or in the worst of circumstances even going to jail. It's why we have higher-ups sign off on things. The signature of a higher-up means that your ass is covered because they've taken a step forward in the blame-taking line. And to be fair, covering your ass isn't the dumbest thing to do when many of those same organizations live on the blame model themselves. When things go wrong, your boss looks for a scapegoat. You want to make sure it's not you.

An eye for an eye and now everyone is blind. But there's another way. What if, instead of being brought up with and working in cultures of blame, we were brought up with and worked in cultures of opportunity? Fire up the kumbaya machine!

I BLAME CORY DOCTOROW

I've been a fan of Cory Doctorow's writing since Down & Out In the Magic Kingdom, both his fiction and non-fiction. He's a very good writer and spins a hell of a yarn. A couple of years ago, I was reading Walkaway, which had just been released, and something jumped out at me. It's a great book and I encourage you to read it, so I'll try to explain this without revealing too much of the plot. The book is set in a dystopian near-future (as opposed to our dystopian present) where some people opt out of society and go off into the woods to set up their own communities. These communities are egalitarian by design, so there's no one around to assign work or to level blame. When someone notices something that needs to be done, they either do it or throw it on a job board. If you're looking for something to do, you go to the job board and pick a job and do it. But what if you slack off and do a crappy job of it? Well, the cycle just repeats. Someone notices it needs to be done and either does it or throws it on the job board.

So here's the thing. I'm reading along, thoroughly enjoying this book and I got to that part and my core seized up. What. The actual. Fuck. Someone just got away with doing a shitty job. Hold up. Someone needs to be blamed for doing a shitty job here, Cory! We can't get away with letting someone do a shitty job! Honestly, it made me angry. (I just laughed out loud writing that last sentence, by the way.) Nonetheless, I kept reading. And eventually got the point.

The important thing was that the job got done. Yes, the first person who took the job had the opportunity to do the job well. And yes, they should have. But bringing the community to a halt to assign blame and prescribe punishment wasn't bringing the job closer to being done well, either. If anything, it was scaring off anyone else who might've been willing to take the job. If the goal was to get the job done the thing that made the most sense was to give someone else an opportunity to do it.

Reader, I called my mother when I figured this out!

"Hey, Mom, it's been a while."

"And whose fault is that?"

LET'S GO FOR A DRIVE

I grew up before GPS systems. And with a father who refused to read a map. This meant that any trip to a place we weren't familiar with was an opportunity for sacramental display, with my mother, who couldn't read English, being required to navigate us to the right place.

"We missed the turn!"

"And whose fault is that!?"

Trips were exhausting. And quite often involved crying.

I don't drive much, what with living in a small city with semi-functioning public transportation. But every once in a while I have to rent a car. For example, a few weeks ago my wife hiked a portion of the John Muir trail with some friends and I had to rent a car to pick her up at Yosemite. I didn't know the route, so I turned on Google Maps and got turn-by-turn directions. Here's the amazing thing about Google Maps: you can miss as many turns as you want, and it never yells at you or blames you for fucking up. It just reroutes—it looks for an opportunity to get the job done.

If we can build software that focuses on opportunity instead of blame, surely we can build workplaces that do the same, right? Maybe? I hope so.

SO EVERYONE'S JUST GONNA FUCK EVERYTHING UP AND I HAVE TO PICK UP ALL THE PIECES?

Does this mean we shouldn't hold people accountable when they fuck up? Absolutely not. But it does mean that having someone else to blame or being able to cover your own ass shouldn't be the standard we aspire to. Yes, your product manager may have more power in the organization than you do. Yes, decisions may be getting made above your pay grade. All that may be true. But the work is still coming through you. You have an opportunity to do good work. Take it.

Does this mean you're gonna end up fixing a lot of things that other people screwed up? Maybe, but the important part of that sentence is that those things are now fixed! The most important thing about bad work isn't whose fault it is that the work is bad. It's that it is bad work. The most important thing to do is to fix that bad work.

Assigning blame might be a useful way to keep bad work from happening again, but that's more akin to finding a problem in the production line and rectifying it. It would be stupid not to do that.

When assigning blame becomes a survival tool to make sure that our ass is covered and we're not the ones being scapegoated, then it's a useless tool and it points to a problem in the culture. We behave like this because the system rewards us for behaving like this. And ultimately, bad work is everybody's fault. Good work means everyone availed themselves of the opportunity.

WHO DO DESIGNERS REALLY WORK FOR

(Originally published in Adobe Xd Ideas on September 20, 2019)

I've been working in client services for almost twenty years. That's long enough to learn a few things. One of the things we learned along the way was that our clients (this goes for bosses as well) need to know who exactly they are hiring and what it's going to be like to work with us before we all actually agree to work together. And because we've had one too many arguments that ended with "I sign your paychecks and you will do what I say," we came up with a little thing we tell all our clients before they agree to work with us:

"You may be hiring us, and that may be your name on the check, but we do not work for you. We're coming in to solve a problem, because we believe it needs to be solved, and it's worth solving. But we work for the people being affected by that problem. Our job is to look out for them because they're not in the room. And we will under no circumstances design anything that puts those people at risk."

Ballsy, eh? Only a few people have refused to move forward after hearing that. And trust me, I say it in a nice way. I'm a people person. But anyone who refused to work with us after hearing that was doing us a favor; they were probably going to be a nightmare client. More often than you'd expect, the reply we got was "Awesome. That's exactly what I want."

And here's the important thing: I absolutely believe every word of what I was telling them. When you hire me as a designer, I do not work for you. I may practice my craft at your service, but you haven't earned the right to shape how I practice that craft. One, you don't want me designing at your level, you want me designing at mine, which means you don't get to pull the strings[1].

---

1 *For the record, this doesn't mean I don't need your feedback or collaboration. I do. It just means I'm doing the driving, which is what you're paying me for.*

I do. Two, you're hiring someone who performs a service, not a servant. There's a difference. I'm not there to do your bidding, I'm there to solve a problem or reach a goal that we agreed upon.

So who do designers really work for?

DESIGN ETHICS AND THE HIPPOCRATIC OATH

In the last few years, I've developed a useful little trick; I look at other professions to see how they behave in certain situations, and then attempt to map them over to what we do. This is helpful because it gives us as designers a little distance, and allows us to learn from people who've already solved similar problems. Or, as my mom would say when she took my brothers and I out to dinner and we behaved like assholes, "You see that table over there? You see how they're not throwing food at each other? Their parents won't be divorced in a year. And their kids will grow up to be doctors." This may be why I write about doctors so much, by the way.

Doctors take an oath before they begin practicing. This doesn't ensure they're all going to behave ethically, but if they're going to behave badly, they certainly can't claim ignorance. Now once they take that oath, they can go off and do a variety of things. Some enter private practice. Some join organizations like Medecins Sans Frontieres. Some go to work in top-of-the-line hospitals serving patients with lots of insurance. Some go to work at free clinics. A lot of them end up doing a combination of those things. But no matter where they go, the oath they took determines how they behave on the job. They'll certainly face constraints along the way, such as the hospital they're working on not having the latest equipment. But their job is to do their job, as defined by their code, to the best of their abilities.

Pay attention, because this is where the comparison goes into high gear. Now imagine a doctor runs into a sketchy hospital administrator who's trying to keep a hospital afloat by doing things like telling them to order tests patients don't need, or prescribe medications from pharmaceutical companies that he's made deals with, or charging people for private rooms they didn't have... you get the idea. This isn't much different than working for a boss who asks you to target addictive products at poor people, or to get user data you don't need in case they might want to sell it later. The difference is, when a doctor is asked to do those things, the oath they took supersedes the signature on their paycheck. When a designer is asked to do those things, there's no oath in place. No ethical framework to fall back on. You may get a gut feeling that what you're doing is wrong, it may not feel good to do it, but at no point in your career have you actually put pen to paper, or hand over heart and promised not to behave this way.

More importantly, if a doctor behaves unethically and is caught there's a fairly good chance they could lose their license. A designer who behaves unethically for a shady boss might get a raise. Your shady boss now knows they have someone they can rely on for shady work. "But people die when doctors do their jobs badly, Mike!"

DESIGN CAN KILL, TOO

In 2017, the Royal Society of Public Health, in conjunction with the Young Health Movement, published a study about social media and mental health for young people. It's worth reading in its entirety, but let me highlight the part that's salient to what we're discussing here. Between 2010 and 2015, after a 20 year decline, teenage suicide started rising again. Along with rates of anxiety, depression, body dysmorphia, etc.

"Social media has been described as more addictive than cigarettes and alcohol, and is now so entrenched in the lives of young people that it is no longer possible to ignore it when talking about young people's mental health issues." —Shirley Cramer, the chief executive of the Royal Society for Public Health

And while the study doesn't make a conclusive connection between these things and social media (because of academic rigor and all that) it makes a very strong case for it. Thankfully, I'm not an academic, and I have little patience for academic guidelines. So I have no problem telling you this: the work we are doing is killing people. A Google search for "deaths by social media" will bring up more examples than you should need.

Those of us who grew up designing things online need to realize the repercussions of the work we do. We're no longer pushing pixels around. We're building complex systems that touch people's lives, affect their personal relationships, broadcast both words of support and hate, and undeniably affect their mental health. When we do our jobs well we improve people's lives. When we don't, people die.

So, yes. The comparisons to the medical profession are apt.

CELEBRATE (AND DESIGN FOR) DIFFERENCES AND INCONSISTENCIES

I was taking the metro home this evening and realized I'd left my headphones at the office. Which means I had to listen to people. I heard two tech dudes arguing about how to set up a server. I heard two other dudes arguing about data collection. And I watched the dude next to me do some coding. On a 15-inch laptop.

On a crowded metro. When we got to my stop the doors wouldn't open because of some technical malfunction. Everyone waited mostly patiently as the driver got out and opened each door one by one, which wasn't quick. While he did that, the two tech dudes talking about setting up a server changed their topic to how bad San Francisco's public transportation can be (they're not wrong). One mentioned how inefficient the city was. He pointed out the metro stops are all different. Sometimes underground, sometimes above ground. Sometimes there's a platform. Sometimes the steps have to lower to meet the street. Sometimes the doors on the left open. Sometimes the right. The other guy replied that things would certainly run a lot smoother and more efficiently if we standardized all of that.

And he's not wrong.

Society runs more efficiently when all the metro stops are the same. And all the streets are a certain width. And everyone would just agree to behave the same way. And follow the same rules. And eat the same thing. Soylent is very efficient. We could all wear the same shoes. (Count the number of All Birds in the room right now!). What if we all voted the same? And spoke the same language?

When I was a little baby designer I was taught that good design meant simplifying. Keep it clean. Keep it simple. Make the system as efficient as possible. As few templates as possible. I'm sure the same goes for setting up style sheets, servers, and all that other shit we do. My city would run more efficiently if we simplified everything.

But I wouldn't want to live there.

My city is a mess. My country is a mess. The internet is a mess. But in none of those cases is the answer to look for efficiencies, but rather to celebrate the differences. Celebrate the reasons the metro stops aren't all the same. Celebrate the crooked streets. Celebrate the different voices. Celebrate the different food smells. Understand that other people like things you don't. And you might like things they don't. And it's all cool! That's what makes this city, and all cities, a blast. And when all these amazing people, some of them who we don't understand at all, go online they are going to behave as inefficiently in there as they do out there. And that is awesome.

And your job, the glorious job you signed up for when you said you wanted to be a designer, is to support all of these people. Make sure none of these incredible voices get lost. And to fight against those who see that brilliant cacophony as a bug and not the greatest feature of all time.

You are our protection against monsters.

Society doesn't serve Silicon Valley. Silicon Valley needs to serve society. And we are big, and we are multiple, and we are amazingly inefficient. We don't all want the same thing. Except that we actually do. It's to thrive.

BAD WORK IS ALWAYS YOUR FAULT

(Originally published in Modus on June 26, 2019)

> Q: I work at a large company. Decisions are made way above my pay grade. I'm able to influence small things, but not the larger picture, even when I know decisions we're making are unethical—such as using what I know to be a dark pattern to keep our users from unsubscribing. I certainly don't want to get fired. If the decision is coming down from management, can I even be held responsible?

I completely understand how hard it is to get a job. I don't want you to get fired either. You probably worked hard to get where you are. You probably went through multiple interviews. You probably got asked a ton of questions by a ton of people. You might have even taken some sort of test to get the job. And I bet there were other people applying for it as well. So congratulations on making it through all of those hoops. You got a job as a designer.

Now you gotta do that job.

The job of a designer is complex. It's a profession, and like most professions it gets defined by the profession itself, not by the people who hire you. For example, if I'm an accountant and I get hired to be an accountant, the expectation is that I'll be doing accounting—numbers and shit. Now, there's a right way and a wrong way to do accounting. The right way is to do the math and let the numbers tell the story. The wrong way is to figure out the story you want to tell, and then make the numbers match the story. In other words, there's an ethical component to it. Fudging the numbers also happens to be illegal, because it has tax implications, and most nations have a robust set of tax laws against that sort of thing.

Does this keep accountants from cooking the books? Not as much as you'd hope. But because accountants are licensed, and tax cops like to look over company books, there's a price to pay for getting caught. That price can range from losing your license to going to jail. And trust me: almost every accountant facing jail time told a judge the decision to cook the books came down from management.

Even once you're out of jail, it's going to be hard to get that license back. Standards of conduct are set by the profession and its licensing body, when there is one. Designers don't have one—yet. We can, however, go to jail now!

If you've read my book, *Ruined by Design*, you're familiar with the story of James Liang. He was the Volkswagen engineer who designed that software that lied about Volkswagen emissions. He was convicted of fraud for knowingly designing software that deceived people. He's currently serving a 40-month sentence for that. And yes, James told the judge the orders came from management. We now have precedent for deceptive design sending people to jail.

You may also be familiar with a tool that Uber designed a few years back called Greyball. In short, Greyball was a hack that Uber tied to the phone numbers of journalists, government regulators, civil servants, and anyone else who was on Travis Kalanick's enemies list. Anytime Uber got a request from a number on that list, the user was told there was no car available. So why aren't the people who designed that sharing a cell with James Liang? That's a fantastic question.

Uber lobbied long and hard to be labeled a "software company" rather than a taxi company or an automotive company. That wasn't about prestige. It was because those other types of companies are regulated to the gills. Software companies? Not so much. But that's coming. Regulation comes when an industry starts hurting people. For example, if you're a Facebook employee, you might notice a poster hanging in your campus that says "Orville Wright didn't have a license." Which is true. It's hard to get a license for something you're inventing, which might be Facebook's point. They're inventing things. Good for them. But when the things they're inventing attract two billion users, and those things start breaking and people start getting hurt, regulation comes. Orville Wright didn't have a license; his was the only plane in the sky. But none of you would get on a plane with an unlicensed pilot at this point. And we're super glad there's a whole bunch of air traffic controllers helping those planes not hit each other. Licensing and regulation are signs of a mature industry.

Digital technology is now a mature industry, which means we need to behave responsibly and take our jobs seriously.

Let's get back to you. When you got that job as a designer (congratulations again, by the way), you were given the responsibility to do the job the right way. Regardless of the pressure you might be feeling from the people around you to do it the wrong way. Your primary responsibility is to the people who use your product, much like a pilot's primary responsibility is to their passengers. I'm writing this on a plane, by the way, and the idea that there's someone in that cockpit whose main priority is my safety helps me not freak out that I'm in a metal tube hurtling through the sky.

So, think of the people using your product or service. They're giving you their data. They're entering sensitive information like credit card numbers, bank info, mailing addresses, their kid's pictures. And they're doing this because they believe there's someone on the other side of that screen who's looking out for them and making their safety a priority. Like my pilot.

BUT IT'S NOT MY FAULT

Let's talk about fault, because there's a lot of that going around. Design Twitter is buzzing with designers talking about how their company's bad decisions aren't their fault. First off, you have a responsibility to do the best work possible and to make sure the people on the other end of that screen are getting the best experience they possibly can. So, when you see a problem and let it slip by because your ass is covered and you'll be able to point to somebody else and say it's their fault, what you're really saying is that you care more about what happens to you than what happens to the person on the other end of the screen. That's not okay.

Imagine my pilot again. Let's say that before we took off, one of the maintenance workers missed a step. Now we're in the air and having engine trouble. Not the pilot's fault! So they come running down the aisle strapping on a parachute and screaming, "Not my fault!" before decompressing the cabin and saving their ass. Too dramatic? Sure. And it's also freaking me out. Remember, I'm on a plane right now. The correct way for that pilot to behave when the engine starts misbehaving is to try to bring the plane down in the safest way possible. Their priority is to their passengers. If you're a professional, responsibility trumps fault. Every single time.

SO WHAT DO I DO ABOUT THOSE SKETCHY REQUESTS?

You flip your desk, quit your job, and torch the place!

I'm kidding. Listen, the fact that you're asking this question means you care about this shit, so you're exactly where you need to be. You just need to make your case and stand your ground.

First off, let's not assume they know they're asking you to do something sketchy. Give them the benefit of the doubt. You may be seeing something they didn't when they asked you to do it. You're a designer—you see things they don't. So it's your responsibility to bring it up. If I hired someone to look out for this stuff, and they noticed but didn't bring it to my attention, I'd be pretty annoyed. You weren't hired to mindlessly execute. You were hired to solve problems. That includes finding them and reporting them.

If they do know they're asking you to do something sketchy or unethical—maybe even illegal—let them know. Tell them you cannot do it. Remind them they hired you to be a designer, and that being a designer comes with a set of responsibilities to work ethically, and that you're bound by a code, same as every other professional they hired. Make a comparison to a profession they understand better: "This is like asking the accountant to cook the books." (If they look at each other and smirk, that's a red flag.)

Build alliances around the office. Gather people from other departments together and see if they'll support you. Chances are you won't be the only one who has a problem with what they're asking you to implement. Make the business case: Fooling people might work great in the short term, but it erodes trust in the long term. And none of this even needs to be adversarial. You're a designer giving professional counsel, which is part of the job. So give it in a professional manner.

In the end, if you're working for good people—and let's pray that you are—standing up for yourself and for your users will earn you their respect. Maybe begrudgingly at first. That's okay.

But remember, everything you have a hand in making bears your fingerprints. You're responsible for it. There's no question of that. And if you become the designer who'll do the sketchy work without complaint, you'll become the designer who gets those requests forevermore. And that reputation is hard to shake.

BUT WHAT IF I REFUSE TO DO THE WORK AND THEY FIRE ME?
Well, this is when we start talking about unions. Stay tuned.

WHEN TO PUT DOWN THE TOOLS

(Originally published on Medium on November 14, 2018)

On November 1, 2018, I was teaching an ethics workshop in Ottawa, Canada. One of the topics we covered that day was the power of organized protest. If you walk out of your job in protest, you have a problem. But if you can talk your entire department into walking out with you, then your company has a problem. It's hard to replace an entire department. Especially once people start asking why you're hiring an entire new department. Collectively we have more power than we do individually. The attendees at the workshop might've been buying all of this in theory. But practically? Who's going to do that? Who walks out on their job anymore?

As the workshop drew to a close, I tried to content myself with the fact that I'd gotten people maybe halfway there. Then I glanced at my phone. And saw an alert about 22,000 employees walking out of their jobs at Google all over the world to protest the company's mishandling of a high-profile sexual harassment case. The world noticed. And the workshop attendees who were all packing up to leave noticed. They were all looking at the same, or similar, alerts. They looked at each other. They looked back at me. And I saw hope in their eyes.

What I'd (maybe) convinced them was true in theory, they were now seeing was possible in practice. 22,000 employees at one of the biggest, most powerful, companies in the world, had organized across multiple time zones. They communicated. They planned. They made an impact. And not a single one of them was fired. The story wasn't reported as being about disgruntled employees, which it might have been, had there been three or four employees standing outside a building. It was reported as being about bad corporate policy, which it very much was. Sexual harassment, to be exact.

When we organize, they have a problem.

The biggest companies in the world are made up of people like us.

One of the questions we need to ask ourselves is where the line is. What's the thing that makes you put down the tools. On November 1, 2018, 22,000 Google employees showed us where the line was for them. That walkout brought attention to a situation and it told the company they'd crossed a line. A line those employees wouldn't cross with them.

On November, 14, 2018, The New York Times published a story on how Facebook dealt with the combined crises of Russian meddling, data hacking and hate speech. This is the paragraph that caught my eye:

> *Facebook employed a Republican opposition-research firm to discredit activist protesters, in part by linking them to the liberal financier George Soros. It also tapped its business relationships, persuading a Jewish civil rights group to cast some criticism of the company as anti-Semitic.*

Sadly, it wasn't Facebook's behavior which surprised me. A scorpion does what a scorpion knows how to do. What surprises me is that Facebook employees are still at their desks after finding that their company was actively attempting to discredit activists. No doubt some of them are shook. No doubt some of them will make public statements against their company's policy. And those are needed. No doubt there will be internal spirited conversations within the company. And those are needed as well. But there won't be a walk-out. I say this hours after the article was released. But I doubt that I'll have to come back to this paragraph and revise it. I wish I wasn't so sure of that. But I am.

Facebook employees, with a few individual exceptions, don't believe their company has crossed a line yet. Twitter employees, again with a few individual exceptions, don't believe their company has crossed a line yet. We know this because they haven't put down the tools. And by continuing to aid the companies making those decisions by selling them their labor, they've become complicit in their actions. They haven't organized. They haven't made a stand.

And they won't.

And, believe it or not, that's the hopeful part. Companies where employees aren't taking a stand, companies where employees aren't awake to the complicity of their labor, companies where employees aren't willing to put the tools down to take an ethical stand, will eventually die. They're creating a workforce no one wants to join and building toxic products no one wants to use. Their short-term decisions are digging their long-term resting place.

After watching 22,000 Google employees walk out in protest of their company's unethical actions, I know someone's watching the gate. I'm more inclined to trust them. Not necessarily because I trust the company, but because the employees have shown me that I can trust them.

10 THINGS YOU NEED TO LEARN IN DESIGN SCHOOL IF YOU'RE TIRED OF WASTING YOUR MONEY

(Originally published in Dear Design Student on August 16, 2016)

I hope everyone had a good summer. It's hard to believe it's over, yet here we are. Still feeling the summer breeze of warm ocean wind. Still smelling the sunscreen, the taste of cotton candy, the mind still wandering to that summer crush we'll never see again and sloppy awkward kisses under a summer moon. But as it does every year, and much too soon, summer ends, the Earth tilts, the leaves begin to turn, and we head back to school.

Welcome back, design students. It's time to get back to work. Some of you may be entering your first year of design school at one of our fine design institutions. Some of you are mysteriously returning to finish those degrees. Either way, I wish you the best of luck this semester.

This article is for all of you. I want to help you make your design education as good as it can possibly be. This article is for everyone who's ever said "I wish I'd learned that in school." And while it's certainly not exhaustive, I've compiled a few things that you should make sure you're learning in school.

Also, these aren't things that you need to be in school to learn. But for those of you who are there, might as well learn them early. You'll have a leg up on everyone.

Let's start at the top, shall we:

1. DESIGN EDUCATION IS BROKEN

So here's the deal. I'm not going to lie to you. Design education, with a few exceptions (you can pretend your school is in the exception if it makes you feel better about that fat check you wrote), is inherently broken.

How do I know this? Good question. I run a design studio. I interview design school graduates for a living. And while I see many talented people, and a few people with the potential to have careers as designers, almost every single person I've interviewed lacks the basic core of what it takes to become a professional designer coming out of school. This isn't their fault. Their school didn't teach them what they needed.

This is how I want to help you. I want to help you have a successful design career. Which has nothing to do with how creative you are. I've seen plenty of creative people's careers derail because they couldn't manage their shit. I want you to manage your shit. Realize right now that no one is going to be so dazzled by your work that you don't need to pay attention to this stuff. Because no matter how good your shit is, at some point you are going to have to sell it, invoice for it, and collect your money. Those are the things that prolong your career.

Your school probably has something called a portfolio class. Or a professional services class. It's the class you don't want to take. When I was your age I didn't want to take it either. And it's probably taught by someone who drew the short straw. They want to teach it as much as you want to take it. Thing is, you gotta take it. And you have to demand that it's taught correctly. So feel free to show this list to that professor. Feel free to ask them if these things are in the class curriculum. Tell them these are things you want to learn.

What if they tell you this stuff isn't important? One, they're lying. Two, get out of there. You're wasting your money. And don't let them tell you these are "soft skills."

Being able to pay your rent is not a soft skill. It's a life skill.

2. DESIGN IS A BUSINESS SKILL

You're probably in the same building as the art school kids. Design has as much to do with art as a lobster has to do with a carrot cake. If you truly want a career as a designer, you are going to need to speak about someone's business and organizational goals. You're going to have to learn how to analyze data, you're going to have to learn how measure effectiveness. You're going to have to learn how to build and extend brands. You're going to have to learn how to do goal-driven work. Design is not about expressing yourself. Design is not about following your dream. Design is not about becoming a creative. If that's the kind of stuff that you're interested in that's absolutely fine. Walk across the hall to the art department and learn how many yams you can shove up your ass. (For the record, I have an art degree and that is an excellent reference. Look it up.)

But if you want to be a designer tell your professors you need business skills. If they roll their eyes you're in the wrong school.

3. DESIGN DOES NOT SELL ITSELF

At some point in your design career you are going to have to show your work to a client and convince the client that work is correct. You're going to have to be persuasive. You're going to have to work a room. You're going to have to repeat some of those business goals from the previous point back to them. And you're going to have to explain how your design will achieve those goals. Never, in the history of design, has good design sold itself. Never has a client been blinded by true genius. No, you're not going to be the first. You have to work it.

Tell your professors you need presentation skills. You need practice in front of a room. And it should be a room of people who don't give a shit about design, not your classmates. That's called a critique. We'll get to those in a minute. You need to be able to talk about your work in a way that the people who will hire you understand. No, they don't give a shit about fonts.

4. LEARN HOW TO ASK QUESTIONS

Your first day at the job someone will throw something on your desk. They'll say "Make this." Your reply needs to be about 500 questions. "Why?" "Who is it for?" "How do those people behave?" "Can I talk to some of them?" "What are we trying to achieve?" "How will we measure its success?" etc etc etc. Design is the intentional solution to a problem within a set of constraints. If you don't know what the problem is, you can't design a solution for it. Because you have no idea whether your design would solve it. So the first step in any design exercise is to understand the problem. To do that you have to ask a lot of questions.

Learning what questions to ask, who to ask them of, and how to ask them is a skill that needs to be taught.

5. LEARN HOW TO SAY NO

Throughout your career you will be asked to work for free. You will be asked to work against your own interests. You'll be asked to design things counter to goals. You'll be asked to design according to whims. All those things will fail. And those failures will be on you. As a designer it's on you to do the job to the best of your ability. Learn how to protect yourself by saying no.

Being able to say no without coming across as a petulant asshole is a skill. A skill many people don't have. Ask your professors to teach you how.

6. LEARN HOW TO WRITE AN INVOICE

Your career will be short if you can't get paid. And never accept someone else's terms without question. This also goes for writing statements of work, proposals, and change orders. Change orders can change your life.

7. LEARN HOW TO STEAL

Be aware of your history. Design is the oldest profession in the world. You're not the first person to tackle whatever design problem you're tackling. See how others tackled it. Take the best solutions you find and improve on them. Don't burn time solving things from scratch. Make use of what others have learned.

8. LEARN HOW TO READ AND WRITE AN EMAIL

Your success is going to be determined by how well you can communicate. And possibly a little talent. But seriously, it's about 10% talent. I'm not even kidding. Most of that communication will be done over email. Most email is written badly. On average in every 100 words, five are important. You're going to need how to find those. Clients will write every manner of bad email. (It's not their fault. No one taught them how to do it either.) It's on you to figure out how to read those emails productively. And then reply to them in a productive manner. You're going to need to learn how to write short, succinct to-the-point emails. And you're going to need to learn when the correct reply to an email is actually a phone call.

Tell your professor that designers need to be communication professionals. If they start talking about the AIGA[1] just walk away.

9. DESIGN THE RIGHT THING

Like all professions, there is an ethical component to design. We make things exist! That's pretty powerful. But we also need to be gatekeepers to the things that we allow to exist. And we need to make sure that the things we build are as safe as possible. This means that we need to think about contexts which are not our own, and people who are not like us. (Feel free to define "us" to your particular context.)

A designer's job is to solve problems. And you have a very limited amount of time on Earth. And limited resources. So make sure the problems you take on are worth solving. Your job is not to do what your told. You not only get to ask why, you have to ask why.

Working ethically is a skill. And it's a skill that needs to be

---

1 The AIGA is a professional organization that holds poster contests and ignores Black designers.

taught. It's not easy to tell the CEO of a Fortune 500 company that the product they just asked you to design is harmful. And it takes more than guts. It takes training. Make sure that your professors are giving you that training.

10. LEARN TO BE CRITICAL

Throughout your career people will tell you your work is bad. They will be right a majority of the time. Doesn't matter how good a designer you are. You're going to have to learn how to take and dish out criticism. There's a respectful way to do it—focus on the the work. But if you withhold criticism because you don't want to hurt someone's feelings you're doing them a disservice as a designer. More importantly, you're doing the people who eventually come into contact with that designer's work a disservice.

So learn how to kick the tires on an idea. Good ideas can take it. And they're made better by it. Look for the holes so you can plug them. Be honest with each other when work doesn't meet the standard of excellence. Do it fairly, do it kindly, but do it. The most unkind thing you can do to another designer is to withhold your criticism.

But Mike, I want to be a designer!

I know you do. And I bet you'll be good at it. I bet you're very talented. Even very creative. You've also decided to make a career of it. For which I applaud you. We need good designers. But we need designers who know how to deal with their shit so they can earn an actual living.

This is what you need to be learning in design school. You need to come out of there ready to enter a professional service. Obviously, this is stuff you should be learning in school. But school is only as good as you want it to be. You can fuck off and get a degree. (I did.) Or you can hold your teachers' feet to the fire and say "I'm paying a lot to be here. And I'm willing to put in the work. And these are the things I need to learn."

Welcome back to school.[2]

---

[2] Here's a point I wish I'd made in the original article: don't go into a ridiculous amount of debt to get a design degree. Graduating with a ton of debt severely limits the type of work you can do when you graduate. You'll be forced to make decisions based on who's offering the biggest salary. And those tend to be large tech companies doing unethical crap. You'll be surrounded by assholes doing assholes' work. This is one of the many reasons we need to cancel student debt, and fund free education in this country. Until we do that, don't worry about going to a "prestige" school. In twenty plus years of hiring designers I couldn't tell you where any of them went to school. It never factored in my decision to hire or not hire someone. You'll find better teachers in state schools anyway.

GETTING YOUR FIRST DESIGN JOB

(Originally published in Modus on May 16, 2019)

Q: *I just graduated from design school and landed my first real professional job as a designer. What do I need to know that I didn't learn in school?*

First of all, congratulations on landing your first job. That's a big deal. You should be proud of yourself. I've got a few tips that might help you out on your first day and beyond. Read on!

IMPOSTER SYNDROME IS EXPENSIVE

One of the biggest issues with designers (all of whom I love dearly) is being saddled with a crippling disease called "imposter syndrome." I've seen it in new designers, and I've seen it in experienced designers. In a later column I'll go over the reasons why this happens. But right now, I'm gonna do that thing you're always wishing your therapist would do, and just show you how to solve the problem.

My guess is it wasn't easy to get the job. There were probably a couple of phone calls and possibly even some stupid test you had to pass before you even got to the interview phase. You probably had multiple interviews, yes? With multiple people in them? Possibly (hopefully) from a broad cross-section of teams in the company. You probably talked to an exhausting number of people for an exhausting amount of time. Including a few who were wearing Allbirds and Patagonia fleece vests, which means you got interviewed by managers. Which, fashion aside, is usually a good sign. It's also a safe assumption that the company chose some of their brightest people to be part of the interview process. They generally keep the dummies in the back.

I'm also gonna go out on a limb here and assume that you weren't the only person who went through this process. They probably interviewed more than a few people for the position you won. So you were one of a few who went through a tedious, mind-numbingly long process to get the job you eventually got.

OK, here's the important part: There are two possibilities here. Either you were able to pull the wool over everyone's eyes and trick them into thinking you knew what you were doing, or... OR... you're actually as good as all of those people thought you were. Which of those seems more reasonable? That's right. The latter. You're actually as good as all of those people thought you were.

You're not an imposter. You got hired because you are good at what you do. Don't forget that. You're welcome.

FIND OUT WHAT PEOPLE DO

You're gonna meet a lot of people on your first week. Be fascinated with them! You're a pair of ears with feet. Ask them what they do. Listen to what they say. You're all working together on the same thing (hopefully), and you're going to collaborate with these people a lot during the job. Show some interest in who they are and what they do. Learn as many names as possible. You want to impress your new co-workers? Then listen to what they have to say. They've been there longer and they can teach you a lot more when they're talking than when you're talking.

UNDERSTAND WHERE THE MONEY COMES FROM

How does your company make money? This is a trick question. Because you really should've found this out before you took the job. But a lot of companies change how they make money fairly routinely, especially startups. Companies need to make money to stay in business. And you're going to be helping them do that. Know where it comes from and know what your role is in making it happen. And for the love of society, make sure that it's happening in an ethical way. You can always check with whoever's in charge of ethics. Which reminds me...

YOU ARE IN CHARGE OF ETHICS

The work you do will affect people's lives. You need to be more concerned with the consequences of your work than the cleverness of your ideas or the profits of your company. Everything that passes through your hands is an opportunity for an ethics check. It's also a responsibility. And yes, I'd be saying this same exact thing to anyone in your company. You're all in charge of ethics.

There are two words that every designer needs to be comfortable saying: "why" and "no." With every project that crosses your desk, you need to ask "Why are we doing this?" Asking that question is part of the job. "Because I am telling you to" is not an acceptable answer, and if there's no answer beyond that, you need to reply with "no."

Design is the solution to a problem. It's your job to fully understand the problem. You ask why until you fully understand what the problem is. Wanting to squeeze more money out of your customers without improving their lives in some fashion is not a problem. It's a character flaw. You cannot do that. You'll have to say no. Will there be repercussions? Yes. But one of those might be the people who hired you respecting you for taking a stand. That's how you know you're working at the right place.

Don't work for people who trick others. Life is too short and the world is messed up enough.

YOU WERE NOT HIRED TO MAKE PEOPLE HAPPY

You got hired because they needed someone with a set of skills that you have. Those skills include not just your labor, but also your counsel. You got hired to solve problems. That means solving the problems to the best of your ability. Not to the best of someone else's ability. Collaboration is awesome and necessary and smart, and it's very different from being an order taker.

Carol in Accounting runs the numbers because it's her job. She doesn't ask anybody if they're happy with the numbers. You solve design problems because it's your job. Don't be asking people if they're happy with your solution. Ask them to collaborate on your solution. Ask them to double-check your solution. Ask them to test your solution. Ask them to break your solution. But do not ask them if they're happy with it.

If you're going to work somewhere that has an already established design practice, you should know what it is. Optimally you knew that before you took the job. If you're going somewhere to establish a design practice, make sure you know how to do that and make sure you have leadership's support. That means you get to actually establish the practice, so if Chip from Sales wants to see the work you're presenting to the team before you present it, and you don't think that's a good idea, then Chip from Sales doesn't get to see the work.

And remember, you are a stakeholder at the company, as much as anyone is. You are an equal.

YOU WORK FOR THE PEOPLE

It's super great that these people hired you, and I hope they're paying you well. But keep in mind that you don't work for the people signing your checks. This may be counterintuitive. But it's gonna come in handy when Brad from Marketing asks you to implement some shady-ass dark pattern on the site. You're going to tell Brad (who'll be wearing a fleece vest, BTW) that you're not going to implement that dark pattern. Brad will remind you that you work for him. And you'll know this isn't true. Because the people you actually work for are the people who'd be tricked by Brad's dark pattern. They're the people you work for. Protect them.

OH, ONE MORE THING...

If this is a tech job, you'll notice that most of the workforce is white dudes. This sucks. It limits the points of view in the office—which means your design solutions are weaker, people who are not white dudes will not want to work there, and you'll see a lot more fleece vests.

If you are not a white dude, congratulations on getting a job in tech despite everything being stacked against you. We desperately need you.

If you are a white dude, I need you to do me a favor. If you're in a meeting and Maria is talking and Kevin from Engineering interrupts her, I want you to turn to Kevin and say "Shut the fuck up, Kevin. I want to hear what Maria has to say."

This not only allows us to hear Maria's idea, but it also increases the chances that Maria will encourage her non-white-dude friends to apply at the company. And that's a good thing.

OUR PRIMARY CONTRACT

(Originally published in Modus on May 29, 2019)

Q: I work at a large social media platform. Last week a doctored video of a civil servant went viral on our platform. Our leadership acknowledged the video was doctored but refused to take it down. I don't think that's the right decision, and I'm not sure what to do. Do I fight or do I do my job?

You do your job. And you fight.

Your job, from the moment you became a designer, was to fight for the people who come in contact with your work. Your job is to help those people accomplish what they're trying to accomplish, whether they're performing a financial transaction, reading the news, watching cat videos, or interacting with other folks online. Your job is to help them do this in as honest a way as possible. No trickery: The cat video is a cat video. The person they think they're talking to is the person they're talking to. And the news they're reading or watching is actually news. They put their trust in your hands, and you need to respect the hell out of that. No bullshit. No misinformation.

Designers have a contract with the people who come in contact with their work: They can trust us not to trick them, not to lie to them, to treat them with respect, and to honor their intentions. If they came to our platform to interact with their friends, we will put them through to their friends. If they came to catch up on the news, we will show them actual news. Our primary responsibility is to the people who use the products and services we build, not to the people who pay us to build them.

Our job is to honor users' intent. Even if their intent is counterproductive to your company's business model. If someone wants to unsubscribe from our platform, as much as that might hurt us, we should make that process as easy as possible for them.

A company that can't survive without tricking people doesn't deserve to survive. And it certainly doesn't deserve your labor.

Nowhere is honoring users' intent more important than when people come to our platforms to be informed on the news of the day. A doctored video doesn't honor that intent. What we're showing them isn't what we claim to be showing them. It's a lie, and we know it's a lie. So we're violating their trust.

So what do we do? Flip tables, pull plugs, and burn things to the ground? Yes! Well, yes, but let's slow our roll just a little bit. There are a few steps to hit before we start lighting molotovs. And they're part of doing the job right as well.

PERSUASION

Step one is to make your case. You were hired at this company to practice your craft, and that includes giving counsel. An accountant's job isn't just adding up the numbers, but also making sure the numbers are honest, and calling bullshit when they're not. Same goes for you. Your job isn't just to make things, but also to be vigilant that the things you make aren't used to harm others. Deceiving people is harmful. And if you think spreading misinformation isn't harmful, I'll remind you that we're currently experiencing the largest measles outbreak in twenty-five years.[1] (You may want to check your company's role in that as well.)

Start with the assumption that the people you work for might not be aware that what they're doing is harmful. After all, this is the kind of thing they hired you to look out for.

"Hey Mark [just a random name I'm making up], we can't be posting a doctored video of a civil servant on our platform. People trust us as a source of news and shit like this erodes that trust. Let's take it down."

"But the video is making us a lot of money" is something Mark might say or get one of his lackeys to say.

"Sure, it's making us money today, but if money is your main motivation, we should look at the financial repercussions of a decision like this over a long period of time. Is a week-long bump in engagement worth the long-term effects of an erosion of trust? But even more importantly than the money, Mark—this is dishonest. We have a contract with our users. I can't break it."

Maybe Mark's motivations are even bigger than that. Maybe Mark doesn't want to take down the video because the company gets $12 million a year in advertising from people who want the video to stay up. Or maybe Mark is afraid the same people who

---

1 *"Hello, this is 2020 calling. About that measles outbreak... hold my beer."*

want the video to stay up might create regulations and laws that would limit, or possibly break up, his business. In which case Mark is thinking long-term. But only about himself. He doesn't care about the long-term effects on others.

In which case Mark (again, just a random name I'm making up) is a real prick, and you're going to have to step up your game. Because you have a contract to protect the people Mark is willing to fuck over to either get (more) rich, or to save his own skin.

That means we need to move to step two: making alliances.

ALLIANCES

You can't be the only person in your company who feels you shouldn't be hosting deceitful content, can you? (If you are, you may want to skip to step three.) Find those other people. Hopefully, you work at a company where employees can get together and discuss things like this in an open manner.

Then, instead of one person bringing up this issue, you've got a group. It's easy to brush one person off. They'll tell you it's just your opinion. (Technically, that may be true, but it's also your expert opinion. Which they're paying you to give.) But it's much harder to ignore a group, especially when the people in that group meet a very important criterion: They're absolutely crucial to the company getting shit done. So the company will probably listen. Or, at least, they'll pretend to hear you out.

Let me tell you a story: A few months ago Microsoft employees found out that ICE was using facial recognition software they had designed to round up immigrants. They weren't having it. They got together, they made a list of demands, and they presented that list to leadership. Microsoft cancelled their contract with ICE. Did they cancel it because it was the right thing to do? Or because they were afraid those employees would bolt? We like to think it was the former, but honestly? It was probably the latter.

It is really, really hard to replace a bunch of employees all at once. Really hard. That's an incredible amount of power, and there's a time to use it. Which brings us to step three: knowing when to put down the tools.

PUT DOWN THE TOOLS

Sadly, we may encounter a situation where leadership can neither be persuaded nor collectively bargained with, which means we will need to reevaluate where we're putting our labor. Are we willing to help someone make a tool which spreads lies and misinformation? Do we feel so defeated that we'll willingly use our skills to harm others? Have we misunderstood the responsibility of our job? Or are we just pricks?

Lord, I hope we're not just pricks. Because I need to believe we have some hope here. We need to be better than the people we work for, because they're showing us who they are every day. And they're not good people.

On December 2, 1964, civil rights activist Mario Savio stood on the steps of Sproul Hall at UC Berkeley and addressed the student body. This is the last paragraph from that speech:

> *"There's a time when the operation of the machine becomes so odious, makes you so sick at heart, that you can't take part! You can't even passively take part! And you've got to put your bodies upon the gears and upon the wheels ... upon the levers, upon all the apparatus, and you've got to make it stop! And you've got to indicate to the people who run it, to the people who own it, that unless you're free, the machine will be prevented from working at all!"*

If the machine is harming the people we promised to protect, it's on us to stop the machine from working. If you claim to be changing things from the inside, at some point you need to see some evidence that it's working. Otherwise you're just banging your head against the wall. And not only is that not doing your head any favors, it's also not really bothering the wall, so consider giving up your seat to someone willing to swing a sledgehammer.

Will there be repercussions? Absolutely. You could miss out on future promotions. Your career could stall out. You could be blacklisted. You could lose your job, which in the United States also means losing your health insurance. But those are the repercussions of doing something. Let's look at what happens when you do nothing: Misinformation spreads. Elections are nefariously influenced. White supremacists come to power. Trans people are stripped of their humanity. Children are put in cages. The list goes on.

Too often when we ask about repercussions what we really mean is "will there be repercussions for me?" But the repercussions of not acting are worse. They may not affect you personally—that's called privilege—but they're most certainly a violation of the contract we made with the people we work for.

We promised we would take care of them. And we promised they could trust us.

I want them to be able to trust us. I hope you do too. But if you don't, I need you out of the way.

HOW CAN YOU TELL YOU'RE BECOMING A BETTER DESIGNER?

(Originally published in Modus on August 8, 2019)

This week's question comes from Twitter, instead of a reader email. I began to reply and then quickly realized it was better answered here, where we have the room for a little more thought and nuance. And this question needs thought and nuance, because it's not as cut and dried as you'd think.

Obviously, you can tell you're getting better as a designer when your craft is a little more polished, when you start thinking more about systems rather one-offs, when you start caring about how someone will move through a page more than how the page looks, when it takes you an hour to do what used to take you all day. These are some of the obvious signs you're getting better as a designer. You can tell the work of a first-year designer from the work of a ten-year veteran. That's the obvious stuff—the maturing-of-talent that comes with getting more experience doing the thing you do. That shouldn't be discounted.

But to really get better as a designer, you have to get better as a person. Because there's a point at which the craft skills hit a wall that can only be climbed by growing as a human being.

Let's start with an admission: When I was in my 20s I thought I was a fucking great designer. I thought I had the answers to everything. I thought I could solve everything by myself; I thought I was a genius. The journey to getting better as a designer was a journey toward understanding how much of an idiot I actually am. Now that I'm in my 50s I understand how little I actually know, how much better the people around me are at a lot of things, and how much help I need from those around me. The journey toward getting better as a designer means being okay with all of those things.

I used to have panic attacks thinking about a boss or a client possibly asking me a question I didn't have an answer for, or asking me to use a skillset I didn't have yet. Now I get excited about the opportunity to either learn a new skill or work with a colleague who already has it. Getting better as a designer means valuing curiosity more than mastery and embracing what you don't know more than fearing it. Especially in an industry where things change every few years.

Fall in love with what you don't know. Because getting better as a designer starts with understanding that you really don't know much at all, and getting really, really good at it means that the percentage of stuff you think you know gets lower and lower as you grow.

I look forward to all of you realizing you're as ignorant as I know I am. Or as my boy Socrates once said, "All I know is that I know nothing." This is also known as the Socratic Paradox. (Ironically, we used to be sure that Socrates said this, but we're not so sure anymore, since Socrates never wrote anything down and all we have are Plato's meeting notes.)

So without further ado, let's go into some of the lessons you'll need to learn to get better as a designer. And let me just say, I've had to learn all of these the hard way. Some stuck more than others. Some I still struggle with. Some I have to relearn every single time.

MOST OF WHAT YOU MAKE IS TRASH

Strap in—I'm about to use a sports metaphor. In 1941, Ted Williams of the Boston Red Sox finished the season with a .406 batting average. He was the first player in MLB to finish the season with a batting average over .400 since 1930, and no one has done it since. (In the interest of justice, let me just say that every baseball statistic before racial integration deserves an asterisk, but that's another article—one someone else is better qualified to write.) In short, in 1941, Ted Williams had the most successful offensive season of the modern era. He hit .406. That means he succeeded in getting on base 40% of the time. It also means he failed 60% of the time. For comparisons' sake, a good batting average these days is considered to be .280, more or less. That means that a professional baseball player can fail 72% of the time and still be considered good.

Designers fail way more than that.

Most of what we do is trash. In over twenty years in this business, I don't think I've gotten better at making things; I've just gotten faster at realizing when I was making trash. I can still recall entire days of laboring over one idea trying to figure out how to make it work, or trying to "save" it, all in the name of preserving my ego. I was terrified that I was making trash. But these days, I just assume I'm making trash. And into the trash can it goes. Trash work for the trash god! Every once in a while, I catch myself making something that's not trash, and I'm pleasantly surprised.

Just accept it: Most of what you make is trash.

LISTEN MORE THAN YOU TALK

The odds that you have the answer to everything are zero. The odds that you have better answers than the people around you when you all put your heads together are also zero. So listen to the people around you. And surround yourself with people who are different from you; otherwise you're just listening to yourself, and that's just another version of you talking. Listen to the people you disagree with more than the ones you agree with; listen to the ones who make you uncomfortable.

People have great stories, and if your mouth is moving you'll never get to hear them. "Tell me more about that" is the gateway to learning something you didn't know, and to gaining the respect of the people around you.

"Oh, but Mike! I know what they're going to say next. I'll get major points if I say it before they do!" No, you won't. You'll look like an ass. And what if you're wrong? Never take away someone's opportunity to surprise you.

GET COMFORTABLE ADMITTING YOU DON'T KNOW SOMETHING

No matter how much you prepare, no matter how much you've researched, no matter how smart you actually are, someone, at some point, is going to throw you a question that you have no answer for. The most confident phrase that can come out of your mouth at that point is "I don't know." Let's look at the options: You could lie, but that's a jerk move. You could try to change the subject, but c'mon, they'll see through that. You could hem and haw and hope an answer falls out of the sky, but deus ex machina isn't an odds-friendly strategy.

Telling someone you don't know something tells them you're not willing to lie to them, because if you're going to lie to someone, that's a prime place to do it. And instead, you did the hard thing. You admitted you didn't know. That takes confidence! Also, it opened up an opportunity for learning. Because they're either going to tell you about what you don't know, or tell you to go find out.

I've worked with a bunch of designers over the years, managed more than a few of them. I've explained things to them, I've gone over project briefs, I've given them all feedback. Afterwards

I've asked them if they had any questions. If someone asks me to re-explain something I know they're paying attention; if they tell me it all made sense, I start worrying. Because for one thing, I'm just not that thorough at explaining things—there's no way I got everything on the first pass. For another, most people aren't good enough at understanding things to get it all the first time around. Get comfortable asking for clarification.

STOP BEING AFRAID OF GETTING FIRED

I've been fired more than once—fewer times than the number of digits I'm typing with, but not by much. At the time, all of those firings seemed totally unfair. In hindsight, I can't believe I didn't get fired sooner from a lot of those jobs. I got fired for various reasons. Some of them because I was being a jerk, some of them because I wasn't doing good work, some of them because I refused to do something that didn't feel right, and some of them because I should have never been working there to begin with.

Every time I got fired it felt bad. Every time I got fired I learned something.

Now, I don't feel good about the places where I was being a jerk; in fact, I've since apologized to those bosses. (Running your own company, and hiring your own jerks, is a great way to develop empathy for your past bosses.) But I've never regretted getting fired for standing my ground and refusing to do something that felt sketchy. Jobs come and go, but trading a paycheck for building something that tricks people, or lies to people, or takes advantage of people is something that never goes away. Which brings us to our last point...

START CARING MORE ABOUT PEOPLE THAN PIXELS

When I was a baby designer I was really good at pushing pixels around. Getting them into the right place. Making them all perfectly aligned. Making sure they were all having a nice little elegant conversation on the screen. Making sure they got along. I still enjoy doing it. It's why I became a designer!

But to be a better designer, a good designer, you need to graduate from the pixel. The pixel is the absolute smallest unit of measurement you can design. Get bigger. To be a good designer you need to care about more than the cleverness of your ideas: You need to care about how those ideas affect people. You need to care about the people on the other side of the screen more than you care about what you're putting on the screen.

You need to care how your work affects their world. You have chosen a profession that touches people's lives. And that's an amazing responsibility. And ultimately, the way you can tell you're getting better as a designer is when you stop caring about the stuff you do, and start caring about the stuff they do.

When someone gets to spend an extra hour with their kid or with their dog or with their family, or sitting alone with a book, or at a show, or whatever, because of something you did? That's when you're getting better as a designer.

STARTING A STUDIO

(Originally published in Dear Design Studio on March 21, 2015)

Q: I'm in my final year of university. I've saved up money and am thinking of opening a design studio straight after I graduate. How did you start off when you first opened your studio? Did you have many clients? How did your potential clients find you? And last of all, how long did it take you to build a good client base?

Let me just say how much I admire the gumption and the confidence of wanting to start your own studio right out of the gate. Don't let anyone take that away from you.

Now here comes the hammer: this is a terrible idea.

Let me tell you a story about my mom. My mom grew up in a small town in Portugal. When she was a teenager, my grandmother took her to the town seamstress, who was nice enough to take her on as an apprentice. She spent years learning the basic tricks of the craft. She learned to measure. She learned to cut. She learned to sew. But she also learned what kind of compliment each of the town's grand ladies needed to hear to approve an invoice. She learned how to talk the larger ladies out of horizontal stripes and the busty ladies out of the shiny red fabric. In short, she learned her trade.

As her skills grew, so did her responsibilities. In a few years, she was the seamstress' most trusted employee. And here's where something great happened. The seamstress pulled her aside and said, "Judite, it's time to start your own shop." And she handed my mother a small stack of index cards with the names, phone numbers, and measurements of good clients my mother had been working with. She said, "Take care of these clients and they will bring you their friends and you will never want for work." And she was right.

The Working Angers

If you are serious about a career in design, the absolute best thing you can do right now is to get yourself a job at a studio working for experienced designers who are willing to teach you the parts of the trade you didn't get in school. A good designer understands that part of their role is to teach the next generation.

You'll be getting lessons on finding clients, handling invoices, salesmanship, what to do when a client won't pay, etc. This stuff is invaluable. You'll be exposed to lots of different types of problems and clients at the same time. And the fact that you're asking me about how to find clients means that school hasn't taught you those things. Design schools rarely do, and when they do, design students aren't too keen on taking "the business class."

Go watch how someone else handles it. Listen to their stories. You will not be in charge, but you are also not ready to be. Slowly, but surely, they will hand you the reins, along with a safety net of being there should you stumble.

As far as your questions about getting clients, it's all connected. Your clients come from the relationships you build up over the years. Relationships with other designers, other shops, and former clients. Most of all former clients! You spend your career building those up.

If it were up to me, and someday it will be, no designer would be able to practice without a two-year residency at a design shop working under someone with twenty years' experience.

I love that you want to start your own company, but don't be in a hurry to do it. There's a lot to learn, and you have the time to do it right.

EIGHT REASONS TO TURN DOWN THAT STARTUP JOB

(Originally published in Dear Design Student on August 26, 2015)

> Q: I graduated from school this year and I've been looking for my first job. After interviewing around, I finally got a job offer at a small startup. How do I decide if it's the right offer to take?

This one is easy. Don't take it. You're just starting your career, and a startup is the absolute worst place for you right now. Let me break it down:

1. YOU DON'T KNOW HOW TO BE A DESIGNER YET.

I hate to be the one to tell you this, but I promised I would never lie to you. You have absolutely no idea how to be a designer yet. You might have been the greatest design student at your school, and you still have no idea how to be a designer. At best, you've picked up a very strong set of formal and aesthetic skills which will serve as a foundation to become a designer. But you've never dealt with a client or a boss. You've never had to sell an idea. You've never dealt with having to convince your engineering team why something was important. You've never learned to say no to a bad request. You've never had to gather requirements, and you've most likely never interviewed a user. Your mileage may vary, depending on where you went to school, of course. This isn't me being a jerk, either. At this stage, there's no way you would know these things. But you want to put yourself in a position where you can learn them. And a startup, where you'll most likely have to do all of these things, there probably isn't going to have anyone who can teach you.

2. NO ONE AT THAT STARTUP IS GOING TO TEACH YOU HOW TO BE A BETTER DESIGNER.

You will most likely be on a very small team of designers, all of them with the same experience as you. Maybe one will have been there six months longer, which means he's making more money. And in a world where you have to watch your burn rate, he's getting laid off first. So, he's not teaching you anything. You may be looking at Silicon Valley's new favorite game: Let's hire 200 designers and see who sticks. Which is not unlike when the sea monkey company would send you a thousand sea monkeys, knowing that 900 would die within the first week. Or, you may be the only designer on staff, which means you're either getting tacked to the marketing team or the engineering team. Both of which will see you as a weird "other type" who they'll use to meet their needs. You'll be making buttons and display ads.

3. YOU NEED A MENTOR.

I think I've told you guys about my mom before.[1] She's a seamstress. When she was a teen, her mother (my grandmother) took her to the best seamstress in town. She convinced the seamstress to take her on as an apprentice. Over the years, she taught my mother the trade. She taught her the technical stuff, but she also taught her how to bill, how to properly charge for her work, how to get new clients. In essence, she taught her how to earn a living. And when the day came, the seamstress gave her a few choice clients from her rolodex and told her she was ready to set up shop for herself. This is the kind of relationship you should be looking for.

Humble yourself enough to be an apprentice. Find a mentor. The good ones are hard to find and aren't usually found at startups, which favor young folks without family commitments willing to put in 12 hour days.

4. YOU NEED TO BE GOING WIDER THAN DEEP RIGHT NOW.

The only problem you will learn to solve at a startup is that startup's problem. And it may well indeed be a worthy problem to solve. But right now you need to be learning how to solve a wide

---

1   Yeah, in the previous article, in fact. One of the things I noticed putting this collection together is how often I return to the same bits and rework them. I don't mind that. It's like when you're learning how to cook something. You practice getting better every time you make it. Sometimes I get stuck writing and rather than sit there like a rock staring at a blank page I decide to see if I can refine an old bit I was never quite happy with. I saw someone refer to this as self-plagiarism once. I was happy not to think the way they do, and wondered how absolutely dismal they must be to talk to at parties.

variety of problems for a wide variety of people. You need to be trying different things. Dealing with different types of clients.

And that mentor we talked about? They'll be an invaluable resource in teaching you how to deal with all of those different people. And reminding you that lessons you learned on a project a year ago are applicable to the current project in a way you hadn't thought of.

I run a design studio, and on any typical day we have about a half dozen different projects running through the shop. It's never boring. We learn a little bit about this and a little bit about that. We get to find out about industries we knew nothing about. And we get to design a variety of things. Every once in a while one of our designers enjoys a project so much that she decides she wants to go focus in that area. Which is great. She's put in her time and made her choice. And she can go off and focus in that area knowing that she has a strong general foundation.

5. YOU ARE NOT GOING TO GET RICH.

Most startups fail. That's just the nature of the business. And as I'm sure someone will point out in a comment, most businesses fail. In fact, if we stretch it out far enough, all businesses fail. But startups fail fast. Which means they have to move faster than they might fail. So a startup is taking on more risk. And they'll ask you to share that risk with them. Sometimes that means offering you equity, in lieu of a good salary, but more often than not these days, they'll offer you equity and a large salary.

Do the math. You will be asked to work incredibly long hours. You will most likely need to be available to answer your email at any time of day, and you'll probably be expected to work weekends when asked. More insidiously, you'll be made to feel like you're not a "team player" if you don't dedicate yourself heart and soul to the well-being of someone else's company. And god forbid you have a family. Do the math. Take that annual salary, and break it down by the actual hours you'll spend working. (Ask a few of the other employees how much time they spend working to get an average.) And that equity? Yes, you could be one of the very few who cashes in. And I hope you are. Just know that the percentage of those that do are very very small. And you're betting your career on it.

6. UNLESS YOU THRIVE IN CHAOS, YOU WILL NOT BE COMFORTABLE.

New companies are making it up as they go. This is not a criticism. This is a fact. And there's a certain level of excitement in that. But you're already busy enough trying to figure out what it means to work as a designer. You've got enough chaos to deal with inside your own little 4' radius. Once your own skillset is robust, then go ahead and put yourself in exciting and chaotic environments. But right now, you'd just be another person running around in a burning house.

You know that whole thing about making sure your own oxygen mask is secure before you try to help anyone else? That's what this is about. Before you go putting yourself in a chaotic situation, make sure you're the one who can keep their shit together. That means someone who has seen a lot, and knows what to do when something goes crazy. That's not you yet.

7. THE WORLD NEEDS FIXING, NOT DISRUPTING.

I hate to tell you this, but right now the startup world, or at least the ones making the majority of the noise, have their heads up their own ass and don't realize it stinks. They're solving problems for the top 5% of the population. How can I get poor people to do my chores? How can I get people to drive me around without having to pay them health insurance? How can a drone deliver my toilet paper within 15 minutes while the person who fulfilled my order sits at her desk crying because she's working a 15-hour day and can't take time off to get that lump in her chest looked at. This is known as the service economy. Where entitled white boys figure out how to replicate their private school dorm experience for life.

Don't play that game. As a newly-minted designer, I want you to consider using your skills for the betterment of society. Go find some real problems to solve. We have enough of them. Check out Code for America, or 18F, or US Digital Services. Our craft is a service that should be used to make people's life easier. And especially those who need us most.

8. DON'T BE SOMEBODY'S MONKEY.

Whatever you choose to do, whether you decide to take this startup job or not, I wish you luck. I commend you on entering the workforce. And I hope you take this article, and everything else you read, with a little grain of salt. Trust in your own abilities. Be confident enough to stand by your ideas, and to admit when you are wrong. Look out for your own needs. Learn to say no more than you say yes. Treat people the way you wish they would treat you. And help those that come after you, like those you came before you are now helping you.

GET PAID WHAT YOU'RE WORTH TO SPEAK AT CONFERENCES

(Originally published in Modus on June 12, 2019)

Q: *I've been a designer for a while, I'm good at it, and I have some opinions and expertise I want to share with the community. I've written a few articles and blog posts that've gotten good traction, I've given a few presentations at work, and now I'd like to try my hand at public speaking. I'd also like to get paid for it. Do you have any tips?*

Yes, I have tips. I was late to public speaking myself. Took a while to conquer the inner saboteur that was telling me not to even try, so first off: I commend you on your bravery. Anyone willing to get out in front of a crowd and talk deserves to be commended for their bravery. It's a hard thing to do. It's also labor, so you deserve to get paid for it. How much you get paid depends on a few things, which we'll get into. But let me say this again so it's crystal clear for anyone who doesn't read further:

Speaking is labor. Labor gets paid. Never let someone convince you otherwise.

LET'S DEAL WITH MY FUCKING PRIVILEGE, THOUGH

When getting advice, you gotta consider the source of the advice. In this case, as is the case with everything I write, the source is a white guy. Like I said, I was late to the speaking gig because I had to get past the voice inside telling me I couldn't do it. But it was relatively smooth sailing once I got past that. No one ever questioned whether I belonged on stage because I looked like everyone else on stage. Some of you will have to deal with your inner saboteur only to be reminded (as if you could ever forget) that this industry is still sexist and racist as fuck. Worse yet, those two forces will team up. The voice inside your head will tell you that the forces working systemically to keep you from succeeding are doing it because you're not talented enough. You are. Those voices are lying to you. I don't know much about being on the wrong side of systemic oppression, but I do know something about voices in your head. Don't listen to them.

Now gimme a second to talk to the white boys: Hey Kevin and Chad. Here's what we're gonna do. Every time we get invited to speak somewhere we're going to ask who else is already speaking. We're going to make sure there are women on the schedule. We're going to make sure there are POC on the schedule. We're going to make sure the LGBTQ+ community is on the schedule. And if they're not, we're going to recommend people for the organizers to put on the schedule and we're going to make that a condition of being on the schedule ourselves. Non-negotiable, gentlemen.

But Mike, isn't this a quota system?

No, Chad. "Every slot belongs to a white boy by default" is a quota system. This is just being decent.

NOT EVERY CONFERENCE IS THE SAME

Before we start talking about how much you can get paid to speak in public, let's talk a little bit about the different types of places you can speak. Because it influences how much you can get paid. Educational events, like speaking at a school, generally offer very little, if anything. But if you can spare the time, and the labor, it's great to help the next generation of designers. Then you've got the local orgs, groups with UX[localname] acronyms that do a lot of professional development and member outreach. Some of them can muster a stipend of some sort. Some can't. After that, you've got your professional conference circuit. Some of which are doing very well financially, some of which are struggling. And most of which are on a rollercoaster of doing well one year, and doing lousy the next.

I have zero interest in throwing shade at people who organize conferences. It's really fucking hard work. It takes a set of skills I do not have. The majority of conference organizers work their asses off almost year-round, try to put on a good show, and care about their community. They're also old, like me. The smart ones surround themselves with young people who tell them when they're falling into old habits, outmoded ways of thinking, lazily inviting the same white faces over and over, and not evolving like they need to be. (That was a giant hint, my old people!) Yes, there are a few bad eggs, that's true of anything. But the bigger problem tends to be people who are well-meaning, if a bit clueless. We can help them.

The professional circuit is where you can start getting paid.

DON'T DO THE ORGANIZER'S JOB

Getting paid to speak requires a negotiation. And everyone has a role to play in a negotiation. Your role is to get paid as much as possible. The organizer's role is to pay you as little as possible. Neither of those roles are negative. The more money you make speaking, the more this becomes a feasible use of your time and energy. The less money the organizer pays out, the more they can ensure the continued success of the conference. They're trying to save money on the venue, the food, the nametags, etc. Don't take it personally.

Make sure you're playing the role you're supposed to be playing, which is getting what you need. At no point should you ask a conference organizer what they can afford to pay you. Because if I'm the conference organizer, and I'm doing my job, I'm low-balling you. I'd be an idiot not to. That's my role in the negotiation.

Let's run through a typical scenario. You get an email from a conference organizer inviting you to speak. They tell you a little bit about the conference, such as the theme (if they have one) and the number of expected attendees. They tell you who's spoken there in the past. If they have other speakers booked for this year's event, they might tell you who they are. If it's an international conference they might tell you a little bit about their city or town, including local sites and cuisine. Sometimes, they'll tell you what they're offering you to speak. Sometimes they won't. Sometimes they'll tell you it's a volunteer-run community event being run on a budget. That's the organizer doing their part of the negotiation.

Here's how you do your part: "That sounds amazing. I'm honored that you'd consider me to speak at such a wonderful event. I charge $5,000 for a talk, plus business class travel, and accommodations for five nights. Please let me know if this works and we can schedule a call to go over details."

Let's break that down:

You're stating your cost. You're not asking them what they can afford, you're telling them what you cost. Your job is to get as much money as you can. Does $5,000 sound like a lot? Well, it might be. I know people who get more. Are they that much smarter than you? Maaaaaybe. But also maybe not. Kevin and Chad wouldn't hesitate to ask for it though. Neither would I. And you're probably not going to get that for your first talk, but it's a fine number to shoot for. Can all conferences afford that? No. But this is about you and what you're worth. Don't be afraid to ask for it.

Business class travel? Some champagne socialist you are. True. My rule is that if I have to cross an ocean I'm flying business class. I'm old and I have a bad back. And I spend a lot of time on planes. I'm not willing to spend twelve hours in a coach seat. I'm doing business and it's right there in the name. Can all conferences pay for this? No. But I've made a decision for myself, and I accept those consequences.

Five nights' accommodation? Well, obviously that's a number pulled out of the air. That depends on where you're going, how much of the conference you want to see, and where it's located. But if you're crossing six time zones, get there a day early to reset your internal clock. And if you're going to a foreign city you've never been to before, give yourself an extra day at the end to explore. Accommodations are the cheapest part of what you're asking for and most organizers are excited that you want to explore their city. They'll give you an extra night.

Now, are you gonna get all of these things just because you asked for them? Absolutely not. But you're certainly more likely to than if you hadn't asked! More likely, this is the beginning of a negotiation. The goal of which is that you land on something that's amenable to both parties, and it certainly doesn't have to be adversarial.

THE COST OF A TALK

"You want how much for a forty minute talk?"

Oh, I'll talk for forty minutes for free. All my kid has to do is ask me whether Carvel is better than Baskin-Robbins. But your forty minute talk took a couple months to write, put together slides, practice, revise, practice some more, cajole very patient friends and family to listen to dry runs to make sure it was conference-ready. Then I had to fly to where your conference was and fly back. And all that was an honor to do! I am not complaining. I love doing it! But it takes a lot of labor, and time. This is part of how I earn my living, by the way. Which means I can't lose money while doing it, and breaking even isn't enough. I am here to make money from my labor. The good news is that if I'm good at my thing, and the conference organizer is good at their thing, we both walk away with a bit of cash, and we've provided a solid value for the audience, meaning they've learned a few things that make them better at their thing.

Also, what I'm charging you is a percentage of the total cost I need to recoup for that talk. I'll be spreading that around amongst several conferences. If a conference wants a specific talk written just for them, they need to incur the complete cost of writing a talk from scratch.

THEIR BUSINESS MODEL IS NOT YOUR PROBLEM

Back in the '90s, during the height of home taping, the record industry got together and did a "Home taping is killing music" campaign. They made a logo with a cassette and crossbones, which frankly was pretty cool. Some entrepreneurial punks answered back by stealing the logo, changing the words to "your business model is not my problem," and making stickers and shirts. Their version was way more popular than the record industry's. Plus they copied part of it, which: chef's kiss.

I'm bringing this up because sometimes when you tell an organizer what you charge to speak, they reply with guilt and manipulation. This is not okay. It's not okay to tell you that if they paid you what you're asking no one else would get paid, or they wouldn't make any money, or that the conference would die if they had to pay you, or that they're doing you a favor, or that paying you means they'd have to raise the price of tickets. In short, that's not your problem. They can either afford you, or they can't.

They can, however, try something like "That's a bit more than we were thinking. Would you consider doing it for $3,000?" Now you've got a negotiation going.

No one is doing you a favor by putting you on stage. You aren't their guest. You're there to work. They may be providing you with an opportunity, but it's an opportunity you earned. You're also providing them with content for their event. The exchange is mutual and benefits both sides equally. The only thing you owe the organizer is a good talk. You are an expert, speaking on your expertise, at a professional gathering. Your job is to deliver the content. The organizer's job is to make the business model work.

And as a conference organizer, please don't ask me not to make money so that you can.

ULTIMATELY, THIS IS YOUR CHOICE

We choose to do some things for free; we choose to do some for money. That's a choice. People get to make the choices they believe are right for them. Let's not shame them for it. Ultimately, this is up to you. You want to talk at a place for free? Do it! You want to charge money? Do it. The worst that should happen is they say no. You want to charge one conference, while doing another for free? Totally your call. There are gonna be times when you're not going to want to fly across the country to give a talk no matter how much they offer you, and times where the promise of a free dinner is all it might take to get you on a plane to Barcelona. You get to make these calls. No one gets to give you shit for it.

I'll give you one important caveat though, once you agree to do a thing for a certain price, that's what you agreed to. The time for negotiating ends when you make the agreement.

You're an experienced adult professional with things to tell people. You've put in the work. You've done the labor. If anyone benefits from your labor, you're entitled to your fair share. Go get it.

THIRTEEN WAYS DESIGNERS SCREW UP
CLIENT PRESENTATIONS

(Originally published in Dear Design Student on September 16, 2014)

The hardest part of design is presenting work. You can't even argue about this. I've seen people who did amazing work get up in front of a client and lay eggs. I've also seen people do alright work, and then wrap clients around their little finger. Optimally, you want to do good work and present it well. But I'd rather have a good designer who can present well than a great designer who can't. In fact, I'd argue whether it's possible to be a good designer if you can't present your work to a client. Work that can't be sold is as useless as the designer who can't sell it.

And, no, this is not an additional nice-to-have skill. Presenting is a core design skill.

The first time I presented design to a client, I absolutely choked. I put the work in front of them and stood there like an idiot. It was humiliating. The next time was a little easier. And the time after that, well, you get the idea. I have done every one of the things on this list. I'm sharing them with you in the hopes that they'll spare you a humiliating experience or two. It'll take time.

1. SEEING THE CLIENT AS SOMEONE THEY HAVE TO PLEASE

Your client hired you because you are the expert at what you do. They are the expert at the thing they do. And you have been brought in to add your expertise to the client's expertise to help them accomplish their goal. (If you're presenting work and are still unclear on what that goal is, we have a bigger problem than this article is going to address.) What they didn't hire you to do is to make them happy, or to be their friend. Your decisions should revolve around achieving that goal, not pleasing the client. And while you should do everything in a professional and pleasing manner, never conflate helping the client achieve their goal with making them happy.

They will ask you to do things that run counter, in your expertise, to achieving the goal. Your job is to convince them otherwise. In the end, they will be better served if you see yourself as the expert they believe they hired. And while this may result in some unpleasant conversations during the project, having unpleasant conversations is sometimes part of the job. Doing the wrong thing to avoid an unpleasant conversation doesn't do either of you any favors in the long run.[1]

2. NOT GETTING OFF YOUR ASS

This is your room. Your first job is to inspire confidence. Not just confidence in your work, but also confidence in your client that they hired the right person. Every interaction is an opportunity to reaffirm their decision in hiring you. Get off your ass and lead this meeting. You'll seem more confident if you're standing up. Your voice will carry better. Be the authority on design your client hired. Work the room. Walk to where you're needed. Being on your feet will allow you to walk from person to person as they ask questions, simultaneously making you look more confident and allowing for more intimacy.

It should go without saying that you dressed nicely and your hands are out of your pockets. Now run your presentation, sport.

3. STARTING WITH AN APOLOGY

Do not start the presentation with an apology or disclaimer.

No matter how much more you had hoped to present, by the time you get in that room, whatever you have is exactly the right

---

[1] *Let me tell you a story. It's about the most unpleasant conversation I've ever had. Back when we started our design shop, Mule, we were still getting our sealegs under us and every once in a while we'd end up working with the occasional asshole. This one particular time, we were working with someone who ended up being so hostile and so unpleasantly aggressive that we ended up caving to every bad idea he had. We just gave up. Because we were trying to avoid unpleasant conversations, we ended up designing things we knew wouldn't work. We promised we'd chalk it up to a learning experience and never do it again. About four months after the project launched, this guy showed back up at our door. He was pissed. He told us he'd just finished laying off half his staff because the project we worked on failed. And just as I was about to very defensively tell him why it wasn't our fault, it dawned on me that it was. Those people had been laid off because we'd allowed bad work into the word with our name on it. In order to avoid having unpleasant conversations. Ironically, what followed was the worst conversation I've ever had. And I deserved it. So now, when I find myself in these mid-project unpleasant conversations I'll sit my client down and say, "Bob, let me tell you about the most unpleasant conversation I've ever had." They usually come around. Feel free to steal this story.*

amount of work. Any resetting of expectations should have been handled before the meeting. Obviously, don't do anything that you'll need to apologize for. Like showing up late. Or forgetting an adapter. Or spilling coffee on your new white shirt.

If you're really not prepared for the meeting, then better to cancel it than to waste your clients' time. (You can get away with that exactly once during a project.)

But by the time you are in that room, be ready to present strong and to exude confidence.

4. NOT SETTING THE STAGE PROPERLY

You have gathered all of these busy people together. They probably have other things to do. So let them know why they are in this presentation. Let them know they are a necessary and important part of the conversation. People like feeling needed. And they hate having their time wasted.

Start the meeting by thanking them for their time. Let them know what their role will be. Why they're here. What you'll be showing them. And what kind of participation you need from them. This is your opportunity to make them feel like the experts they are.

5. GIVING THE REAL ESTATE TOUR

Never explain what they can obviously see right in front of them. They can all see the logo on the top left. They can all see the search box. There is absolutely nothing more boring than a designer walking a client down the page, listing all the things they can already see.

Pull up. You don't sell a house by talking about sheetrock. You sell it by getting the buyer to picture themselves in the neighborhood.

Sell the benefits of the work. Sell how the work matches the project's goals. Sell how their new site is going to crush their competitors, and make them all rich beyond their wildest dreams.

And while every decision on that page should have been made with the benefit of data and good research, people are irrational creatures who don't make decisions based on data and research. They make them based on stories. So find your story and tell it.

6. TAKING NOTES

You're too busy giving a presentation to take notes. You're on stage. Ask someone else to take notes for you. And then post them for the client to review after the meeting so you can agree you heard the same thing.

7. READING A SCRIPT

I'm already asleep.

You need to convince your client that you're excited about what you're showing them. Let's be honest here. This is a show. There's a little smoke and mirrors. There's a little Barnum. Not so much that it's a clown show, but enough that you're building up some excitement. Work toward a crescendo. There's little difference between a designer presenting work and a DJ working a crowd. You are selling design.

Have your facts straight. Have your homework done. Have your data at hand. Know why you've made the choices you've made. Have notes nearby if you need to refer to them, but you shouldn't be sitting near your notes anyway. (Remember, you're walking the room.) But work all of these around an exciting narrative. And practice it enough that you know it going in.

Be a scientist when you work, and a snake charmer when you present.

8. GETTING DEFENSIVE

You are not your work and your work is not you. It is not an extension of you, and it is not your personal expression. It is a work product done to meet a client's goals. The client is free to criticize that work, and to tell you whether he believes it has met those goals or not. You are free to disagree with him. And you are expected to be able to make a rational case for those disagreements. But you are not allowed to get all butthurt about it. This is a job.

There's a difference between defending the work, and getting defensive. The latter is personal, it happens when you're seeing the criticism as a reflection of yourself. Guess what, sport? Good people do bad work sometimes. When the client starts critiquing the work, listen to what they are saying. Don't feel like you have to defend all of your decisions then and there. You also don't have to promise them anything then and there. Sometimes it's best to sit on it for a while. It's perfectly fine to say something like "That's interesting feedback. Let me think about it."

9. MENTIONING TYPEFACES

Clients don't care about typefaces. And if they do, they'll ask.

The thing I've heard most often from clients is "I don't know anything about design." (They're wrong, btw.) This is their way of telling you they're uncomfortable. They hate feeling uncomfortable, and you do too. It's on you to get them back into their comfort zone, which is the thing they're experts in—their business. Which is great, because that's something you are not an expert in. It's great to have one in the room. There's already a design expert in the room—you!

So, when presenting the work, talk about it in terms which relate to their business. Talk about how the decisions you made as the design expert match up to the goals of the project. Then your client can judge those as the subject matter they are. But the color, the type, the design shit—you've got that. If you ask them for their opinion on design don't come crying to me when they give it to you, and you're all like, "They don't know anything about design!" They warned you!

10. TALKING ABOUT HOW HARD YOU WORKED

The worst feedback you can get from a client is "Wow. It looks like you worked really hard on this!"

Stop using your work like a time card. If you did it right, it looks like it was effortless. It looks like it's always existed. And the client will probably be irritated that they paid you for 30 hours of work to do something that looks like it took an hour. Which it did. They're just not seeing the 29 hours of bad design that got you to that one hour of good design. And for the love of god, please don't show them those 29 hours of bad design. A presentation is a shitty place for a sausage-making demonstration, and you'll just come across as a defensive, unsure person needing validation.

Sell the fuck out of that one hour of good design—most people can't do ten minutes of it.[2]

11. REACTING TO QUESTIONS AS CHANGE REQUESTS

"Why is search in the top right?"

"I can change it!"

I don't really need to go any further into this one, do I? Just answer the question as asked. You should be able to answer that.

12. NOT GUIDING THE FEEDBACK LOOP

There's only one question worse than "what do you think?" (It's coming up.)

---

2 *This essay eventually turned into a workshop, and this particular point eventually morphed into "Burying the lede." If your presentation is structured like a timeline of your efforts, it's structured to show how hard you worked. It's defensive. You're looking to get points for having interviewed people and arranging lots of post-it notes on a whiteboard. You're trying to impress your client or boss with your hard work. Consider flipping your presentation so the thing they came to see is the first thing they see, and then work backwards to show how you got there. This is what I've come to call the "body on the floor" approach. Like a good mystery novel, a presentation should start with a body on the floor, and then a riveting presentation of how it got there. That's a better narrative.*

Ever hear a designer scream about a client giving them the wrong type of feedback? I have. At which point I ask them if they told the client what kind of feedback they were looking for, and they just pull the panda hat over their head to hide their anger. Most clients have absolutely no idea what kind of feedback you're looking for. That's not surprising. They don't do this every day. They don't have the training that you do. Nor do they need it, because guiding them toward the right type of feedback is part of your job. (Anything that helps you do your job is part of your job.)

During the presentation, feel free to slap your hands together and say "this is the kind of feedback I'm looking for today!" Here are some suggestions for guiding questions:

- Does this reflect your brand?
- Does this reflect your users' needs as we discussed in the research?
- Does this reflect your current ad strategy?

Keep the feedback questions about things that they're the subject matter expert in. I have absolutely no doubt that they'll give you feedback on color and type and all the other stuff you didn't want anyway. Which you should take with a grain of salt. But that other stuff is the feedback you can't move forward without.

Which brings us to the absolute worst question of all:

13. ASKING "DO YOU LIKE IT?"

Dear sweet lord in heaven above and all his angels, you just gave away the farm. They are no longer viewing you as an expert. You are no longer their equal in expertise. You are no longer the person they feel comfortable enough writing a check to. Even if they don't realize it, all of these things just happened. You are now reduced down to a small child showing your dad a picture of the cat and hoping he deems it worthy enough to put on the fridge anchored by his magnetic Las Vegas bottle opener.

The client didn't hire you to make something they liked, and something they like may not be the thing that leads to their success. Do not conflate the two. This point needs to be driven home from the very beginning of the project. And nowhere is this message more undermined than using language that leads them down a subjective path.

...AND ONE WEIRD TRICK THAT YOU WON'T BELIEVE WORKS EVERY TIME.

Learn the client's goddamn name.

HOW TO READ AN EMAIL

(Originally published in Dear Design Student on October 27, 2015)

> Q: I still get anxious whenever I get an email from a client. Especially an email with feedback in it. And the weird thing is—I'm the one who asked for the feedback! But when I open one of those emails, all I see is a giant wall of changes I need to make. So I avoid opening it. What can I do?

Relax. We got this. First off, take a deep breath.

No, I mean that literally. Opening your inbox is the modern office worker's equivalent of opening King Tut's tomb. Your inbox is full of things that have been gathering dust for eons. (At least mine is.) Our inbox contains both treasure and curse. Like a churro stand at the county faire placed next to a port-a-potty. No wonder you avoid it.

But like I said, we got this. I'm gonna teach you how to actually read a client email. Which they should have done in school. For our purposes today, I've taken a few client e-mails and anonymized them for teaching purposes. These snippets bear no resemblance to actual projects, past or present.

READ THE ENTIRE EMAIL FIRST

The first thing we're going to do is read the email in its entirety. Top to bottom. You're not even going to take any notes. You're just going to sit down in your favorite, most comfortable chair and read it all the way through. If you work in a room with other people, you will do this quietly and privately. If you get the urge to read any part of it aloud to your co-workers, swallow it. Just read quietly. When you've finished, take a temperature test. Are you hot? Angry? That's okay. What matters is what comes next. If you're angry, I want you to go for a walk. Got a dog? (You should get a dog.) This is a fine time to take the dog for a walk. If you smoke, go have a cigarette. (And then quit smoking. It's a filthy habit. This includes vaping.) Under no circumstances are you replying to that email or sharing it with your coworkers until you've done the following steps.

ACKNOWLEDGE YOU RECEIVED IT

Send your client a short reply. This is all it says: "Thanks for sending feedback. Reading through it now. Will be in touch as soon as I've read through it a few times to make sure I've got it all."

That's it. That's all it says. It relaxes the person on the other end. Let's them know you're on the case. And it's absolutely true. (Never lie to a client. It's stupid.)

READ THE EMAIL AGAIN

You're going to read the email again, just like you told the client you were going to. At this point, nothing in the email will surprise you, because you've already read it. So you'll be a little calmer.

Here's why that walk was important: you need some empathy right now. You are a trained designer. You know how this stuff works. Your client most likely does not. And it's wrong to expect that they'll give you your feedback exactly how you want them to, even if you give them fantastic feedback guidelines. And you're going to remember two very important things:

- They hired you because you can do something they can't
- You're the pro here

After your second reading you'll probably have co-workers swarming around your desk asking "what does it say?"

Here's your reply: "The usual stuff. Give me a chance to prioritize it. I wanna make sure I'm not misunderstanding it." Again, this has the advantage of being true.

TRANSLATE IT

Here comes the fun part. Guess what, you're reading that email for a third time. Except this time you're going to pull out a notepad. You're going to rewrite the email. You're going to take it apart bit by bit. And put it together the same way a doctor would set a broken bone. Patiently and professionally.

Divide that notebook page in half vertically. Title the left column "feedback" and title the right column "action." The stuff on the right is going to be the basis of the reply you're going to send them. Ready?

Let's take a look at some sample things you might find in a

feedback email:

> We've reviewed the stuff you posted on Monday and compiled our feedback.

Know what the most important word in that sentence is? That's right—"We!" That's a vague word. So write the original sentence in the left column. In the right column write "find out who's included here." You need to check that against your stakeholder list. If anyone's missing, you may be getting their feedback at a later date. You don't want that. So tell them you need all the feedback before you can move on. In fact, go ahead and send that email now. You may have an answer before you finish this exercise. Better yet—call. The phone is your secret weapon. And don't mention anything else. This is all about that list of stakeholders.

Let's move on.

> As discussed, finding a clear way to represent both make and model on product & series pages is essential. Maybe the make and model could appear in place of the related products on the sample product page?

The first sentence here is gold. It shows that they're keeping the goal in mind and holding you to it. This is good. Treasure comments like this. The second sentence is the one that drives designers nuts. But it shouldn't. It drives designers nuts because the client is telling them how to do something. (Ironically, the same designers will also go apeshit when the client "doesn't tell me what they want.") Here's where you need to be empathetic. The client obviously doesn't feel like the goal is being met, and is being nice enough to offer a suggestion, because they feel like it's the right thing to do. Focus on the former, mull over the latter, and if it's the right thing to do—great. Do it. If it's not, then figure out what the right thing is. So, on the action side of your notepad, you write down. "Good idea? Let's test."

> The author photos are too prominent.

Behold the greatest tool in a designer's workshop: Asking why. Unless this was previously discussed, you have no idea why the client thinks the photos are too prominent. Go find out. On the right side of your notepad write "ask why."

> We're still debating the handling of the related articles. I

think it works. Maybe. Was also looking at how this other site does it. Not sure if it works for us. Not sure. Then again, I kinda like how we're handling it.

This comment contains no actual information, request, or feedback. The client is just thinking out loud. Move on. You want to keep an eye on feedback like this though, because it may reveal a larger issue, such as internal disagreement, or maybe you didn't do quite the selling job you think you did. But if you're just seeing a few here and there it's normal.

Bob in Marketing and Janice in Engineering disagree on how we're handling the headers. Can you please show us two different ways so we can vote?

No you cannot. You get a hold of Bob and Janice, as well as the project lead, and you hear them out. You figure out what the disagreement is and you work carefully and methodically to build agreement with the stakeholders. That's problem-solving. But no you're not going to give them taste-tests. You're not running a froyo shop.

We're missing a few items of metadata: ISBN numbers, format, shipping size and weight.

That's actionable. Fix it.

Can we see this with different fonts/colors/shapes/etc?

Why? What's the actual issue with the fonts/colors/shapes/etc you used? You can't just make a change without finding out why they're requesting it. Otherwise, you're back in the froyo shop. And jumping to these requests is what makes you an order-taker, rather than a designer. A designer solves problems. That statement didn't contain a problem. But it appears that there might be a problem, go find out what it is.

Once we're done with these changes we're excited to get it in front of the CEO/Board!

Did you know the CEO/Board were involved? Why are they only seeing it at the end? You're screwed. Don't show them something they didn't have input into. They'll hate that they weren't consulted and can do more damage to finished work than work in progress.

Once you've reviewed the feedback and come up with how to handle each request, you'll need to prioritize it. This is a fine time to call your team together and review it with them. Along with your plan of action. Because now you'll look like you've got it under control. And that's reassuring. And because they can help you prioritize things. They can also check your math.

EMAIL THE CLIENT

Time to email the client back. Make it short. (Always make it short.) Thank them for the feedback. Tell them you've reviewed it with the team. Tell them all the things you're going to change based on the feedback. They'll be happy to see that. Then tell them that you've got questions. List the questions. Give them a deadline for answering them. Offer to jump on the phone to clarify any questions they might have.

AND A VERY SPECIAL TYPE OF FEEDBACK:

<This little detail is great.>
<This little detail is great.>
<This little detail is great.>
<This little detail is great.>
We're concerned the tone isn't quite there.

Christ on a corncob cross. I've actually gotten this format of feedback quite a number of times. I'll admit it used to get me angry. It doesn't anymore. Therapy and experience are a joy to behold. Remember how I told you to read the whole thing a couple of times before reacting? One was to calm down. The other is because people don't necessarily prioritize their thoughts. Either that or they're trying to load you up with good news before dropping the hammer. Your client is misguidedly being kind here. Focus on the kindness, not this misguided part. You got this.

First off, those compliments at the beginning are moot. Because no amount of treating the details is going to fix the problem. You're going to need a serious conversation with the client, because that's a big one.

Send them an email. Here's what it says: "Hey, thanks for being honest about the tone. Let's jump on the phone and discuss that. We'll leave everything else be until we reevaluate that. How's Wednesday at three?"

STOP ADOPTING OTHER PEOPLE'S ANXIETY

(Originally published in Dear Design Student on September 23, 2015)

> Q: I just got off the phone with a client who sounded really angry. He's worried about the project we're working on together running late. He's also worried about how much stuff he has to do before we can launch it. I reassured him it was all going fine, but that seemed to make him even angrier. What can I do? Hurry up; he's on hold.

Take a deep breath. Let me tell you a story.

There's a set of rules we live by at Mule, where I work. We've compiled them over the years, and like all good cautionary tales, there's a scar that goes with each rule. The list is short, and someday I'll tell you all of them. But today I want to talk about Rule One. Rule One has been around for so long we all have a different story of how it came to be. The origins of Rule One have passed into legend. And it's so ingrained in us that we just refer to it as Rule One.

Rule One is simply: "Don't adopt anxiety."

Clients get anxious. You'd probably get anxious in their position too. They've worked their butts off to get a budget approved, or to save up the money to hire you. Their job might be on the line. Their boss might be breathing down their neck. A competitor might be nipping at their heels. Any of these things are enough to wig someone out. And chances are, they're probably dealing with more than one such thing.

Now, if you remember your high school science you'll remember that anxiety is conductive.[1] It wants to travel from one person to another person. And once it sees itself in another person, it feels justified in being in that first person. (I actually have no idea if any of this is right. I spend most of my high school sci-

---

[1] *This is not real science.*

ence classes getting high. Mostly because it was a way to deal with my own anxiety.) To put it another way: once a client becomes anxious, their primary goal becomes to make you anxious, because that justifies their own anxiety. So, if they're freaking out and they can get you to freak out then OMG WE SHOULD ALL BE FREAKING OUT! And then nothing gets done. Or worse, shit gets done but it sucks. Anxiety doesn't produce good work.

The trick to dealing with an anxious client is two-fold. First, remain calm. Nothing good happens otherwise. You are the expert this person hired. Behave it. Imagine slicing your finger open cutting a bagel. You freak out. You wrap it all up. You go to the emergency room. Do you want your doctor to scream when she sees it, or to look at it and very calmly say "Let's take care of that." Be the calm doctor.

Secondly, be empathetic. Now that you've taken control of the situation by remaining calm, you can acknowledge the client's anxiety without adopting it. What the client really wants isn't to make you anxious—it's to be heard. So hear them. And then repeat their concern back to them in the calm voice of a professional. And say something like "I can understand why this is a concern." (This is a horrible time to try to assign blame. If they're blaming you, let it roll off. It may be their fault the project is in a bad spot, but this is the absolute worst time to mention that. That can get handled in a later conversation.)

Then top it off with reassurance. "We're going to look into this."

That's it, right? Not so fast. Because I guarantee you're about to make a big mistake. You're about to start promising them things. It's normal. They are angry. Their voice is loud. You want them to stop. But let's not make project-altering decisions with a person screaming in our face. That sends a horrible message. All you really need to do right now is buy yourself some time to look at their issue in a calm dispassionate manner. Don't promise more work. Don't promise you'll work all night to solve it. Don't promise anything other than looking into it.

"And then what?!?" they'll scream.

"And then we'll know how to handle it." you'll calmly reply.

And then you'll gather your team together to discuss it. Which means that you need to make sure you're not passing along the anxiety to your team. Even if you've properly dealt with the client, they may have been able to leave a little anxiety behind. Anxiety is conductive, remember? That person was probably yelling at you because someone had just finished yelling at them. So before you mention this issue to anyone on your team, make sure you're calm.

Don't repeat the pattern. Go for a walk around the block. Watch some kitten videos. Take a swig of whiskey. Smoke a bowl on the roof. Just make sure the anxiety ends with you.

Lastly, good clients freak out once in a while. But mostly, they're good clients. Hopefully we're talking about an exception in an otherwise healthy relationship. If you find yourself in a client relationship where anxiety is par for the course, that's abuse. Get rid of them. It's no one's job to be treated badly. You'll find more, better clients.

HOW TO PITCH A PROJECT

(Originally published in Dear Design Student on January 20, 2016)

> Q: My partner and I just started our own firm and we're pitching our first big project in a few days. Do you have any tips?

Yes I do. I rock pitches. Everyone who pitches work should think they rock pitches. Otherwise, you're gonna make some weak-ass pitches. And no one wants to sit through those. Including your prospective client.

For those who don't know, a pitch is when you go talk to a prospective client and persuade them to work with you. There are usually several teams pitching the same project. Your first mistake is to believe you're pitching against them. You're not. You're pitching your ability to do the project right.

Also, I've promised never to lie to you so I'll tell you this: pitching work is nerve-wracking. It sucks. You'll start out being terrible at it, but with enough practice, you'll get better. Slowly. Like a turtle. You won't even notice it at first. Then one day you'll be in the middle of a pitch and realize you forgot to throw up that morning—progress isn't always pretty.

Pitches can be subjective as hell, too. There are too many variables. Some of which you have no control over. Like what did your client have for breakfast? Are they worried about something else? Is their brother-in-law in trouble again? Do they have a buddy at a competing agency? You can nail a pitch, and still lose a job for some crazy-making reason. That said, we've also won jobs because we were the buddy at the competing agency. Or because the client got some good news right before we went in. And I've also walked out of pitches thinking I'd blown it only to get a phone call a few days later finding out we'd been awarded the job. So, I figure the factors we can't control or predict all come out even.

That said, I got some tips to help you out. Does following these tips guarantee you'll get the job? Not by a longshot, but they improve your odds of walking out being able to say "I did everything I possibly could to improve my chances."

Go get 'em, tiger!

IT'S NOT ABOUT YOU

It's never about you. No one cares about you. (Except me. I love you dearly.) No one cares how many awards you've gotten. No one cares you won a Webby six years ago. No one cares if you wrote the book on client services. (I know this to be true personally.) People want to know if you understand and can fix their problems. Remember that these people have probably fought tooth and nail to get the budget for this project. They have a pain point and they're shopping around for the right person to help ease their pain. Show them you're that person.

No one wants an ER doctor who spends an hour showing you examples of their past work.

So, tell them who you are. Spend no more than two minutes talking about yourself or your studio in a way that inspires confidence, and make the rest of the conversation about them.

"But Mike, they told us to bring examples of other projects!" That's right. Because they're looking for their new website in your portfolio. So go ahead and show them a couple of recent projects but tie them to problems they have. For example "Here's Acme Novelty. Acme had a workflow issue very similar to the one you described. Here's how we fixed it for them..." They'll feel their pain point melting away.

CONFIDENCE IS SEXY

As much as pitching may make you uncomfortable, you only have to sit through one. The people on the other side of the table have to sit through a bunch. And it's laborious. They'll hear a lot of bad pitches this day. They'll talk to a lot of nervous people. Have some empathy for what they're going through as well. Don't be boring. Be the pitch that doesn't suck. Be the pitch that makes them think they don't have to go through any more pitches. Be the pitch that makes them think they've found a lifelong partner and never have to go through this process again.

Your confidence is for their benefit. And it's contagious.

ASK GOOD QUESTIONS

The key to every good pitch meeting is to get the client talking. They've waited a long time to get this project started.

They've probably suffered under their current site for way too long. And this is the moment they've been waiting for. And not only do you absolutely need to hear what they have to tell you, they'll feel great telling you. It's like the burden is moving from them to you. And once you have it, they won't want it back so they might as well just hire you. Make sure you walk in with a bunch of good questions to ask.

"What kind of site do you want?" is not a good question.

"What impact do you see the new site having on quarterly earnings?" is a good question.

DON'T MINIMIZE THEIR PROBLEM

Here's a good story. We once took a red-eye across the country to pitch a project we were really excited about (Pro tip: never pitch after coming off a red eye). We knew going in they had a fairly complex editorial workflow issue. Fixing that was the main goal of the project. So we partnered with someone who was really fantastic at editorial workflows. During the pitch, they described their problem and we shot back that it would be a piece of cake to fix. And honestly, with the talent that we'd assembled it would have been. But we minimized their problem. They felt like they were pushing a ten-ton rock up a hill every day and we made them feel like it wasn't a big deal. We didn't get the job.

No matter how easy you might think it is to fix someone's problem, remember that their pain is real. Acknowledge it first. Then fix it.

DON'T BRING SPEC WORK

This is one of the most stupid and selfish things you can do as a designer. Stupid because you don't yet understand the problem you're being asked to solve in any meaningful way. Selfish because the pitch meeting now becomes about this golden turd you just dragged in with you. It's supposed to be about the client. But instead of having any meaningful discussion about what the client is trying to accomplish, we have to discuss something that doesn't have anything to do with anything. And now you're being judged on guesswork instead of your problem-solving skills.

"But Mike! People ask for it."

I've gone to plenty of pitch meetings where the client asked me to bring spec work. And I use it as an opportunity to tell them why I didn't bring it, and how it would have been detrimental to what they're trying to do. The truth is that I have no idea how to solve their problem yet. That's going to come from having a lot of conversations with people. And from doing a lot of research. What

I do have is a process I believe in, one that's worked time and time again. And I tell them that anyone who walks in telling them they already know how to solve the problem is lying to them.

And just like that, I've screwed anyone else who might be coming in with spec work that day.

MAKE IT A KICKOFF

The best way to show people what it's like to work with you is to start working. Forget that it's a pitch. Treat it like a kickoff. Let the conversation flow and gently guide it right into interviews. Make the pitch meeting the first step of the discovery process. Someone mentions Sam in engineering? Ask when you can speak to Sam in engineering. Look, your client wants the pitch process to be over as much as you do. They want to get started. So show them what that looks like. And show them that you're able to take the project by the horns. (Just don't overdo it, this technique takes a lot of nuances. It's an advanced skill.)

DON'T MAKE A THING

For the love of god, don't make a special thing to give them. If your company has written a book, great. Give them a copy. Even better, give them a copy of my book: You're My Favorite Client! (Always be selling.) If you have stickers, T-shirts, pens, give them some. But don't hand them a bespoke craft project to remember you by. They'll feel awkward as they throw it in the trash. And then they'll be too guilty to ever want to see you again.

DON'T SWEAT THE DECK

If you've made it all the way through the deck, you're not getting the project. The sooner you're not paying attention to the deck and just having a normal conversation the better your chances of landing the project. So put a few things together. A couple of slides about you. A couple of slides about them. A couple of slides of past work. But don't make the pitch meeting about the deck. The goal isn't to finish the deck. The goal is to get the work.

Oh, and for what it's worth, I don't show any tattoos until after the job is signed. But if your tattoos and/or body jewelry are visible and they have an issue with it? Or if they refuse to acknowledge your pronouns? Fuck 'em.

There's always another client.

GET PAID!

(Originally published in Dear Design Student on July 13, 2015)

> Q: I'm new to freelancing. And I've already started working on my first project! When should I get paid?

Oh boy. The good news is you're getting this lesson early. And that everything is probably going to be okay. Not guaranteed okay. But probably okay. We'll get back to this particular job in just a second, since it's not bleeding out. But first, we're going to make sure this sort of thing never happens again.

In a nutshell: You need to get paid before you start.

Every time. No excuses. And yes, clients will attempt to tell you otherwise, and that's fine. They're just looking out for themselves. But I need you to look out for yourself, too.

What we do is called a specialized service. (Some idiot reading over your shoulder will probably whisper 'bespoke' at this point, but he doesn't know what the word means.) Specialized here means that we design and build things to serve customized solutions. To do a specialized service, I need to procure resources before I start the work. Time, for example, is a resource. And since resources allocated to a specialized project are no longer useful for anything but that specialized solution, they need to be paid for before they are allocated.

Let me give you an example: you can go to IKEA right now and pick up a shelving unit for $100. They already made it. The shelves are a set size. And if you're lucky enough that it's the size you need, you're set. That's what's called a commodity. But if you live in a weird-sized apartment, like I do, and you need a weird-sized shelving unit to hold your very heavy records, you'll find yourself going to a carpenter. The carpenter will measure your space, ask how strong you need the shelves to be, recommend a few different woods for the job, show you a variety of finishes, and

The Working Angers

quote you a price based on your answers. The carpenter will then ask you for 50% of that estimate, and return six weeks later with the perfect shelves. And ask for the remaining 50%.

If he didn't get that 50% up front he'd be putting his own money up for buying wood and other supplies. And taking a huge risk you'd weasel out of the deal after he did the work. He'd be stuck with a shelving unit that only works in your weird apartment. The 50% deposit ensures that we both have skin in the game.

Everyone who does custom work operates in the exact same way. And you will too. And while your customer may not have worked with a designer before, I'm sure they've worked with a tailor, a seamstress, a baker, a wedding photographer, a tattoo artist, a mechanic, a cobbler, an optometrist, etc. So, they know the drill. Your job is to explain to them that you're in that class of professional.

As to the current project: now that you know you should've gotten paid up front, your best bet is to get your client to agree on a milestone for your first payment. Make it one that's coming up pretty soon. And try for 50%, but be ready to fight for it. Your goal here is not to get too deep into resource debt. No one likes to be asked for money when they weren't expecting it, so don't be surprised if your client registers some discomfort with the situation. Remember, you're the one who screwed up. Also, remember that you're under no obligation to work for free. So, if you're getting a feeling that you're going to be left holding the bag, get the hell out.

On your next project, make sure you have a contract that stipulates exactly when and how much you're getting paid. Make sure those milestones are linked to things you control. And make sure you don't allocate your resources until your client has some skin in the game.

But most of all, remember that this is a financial transaction. Design is done for money. Get comfortable with that.

EVERYBODY LEAVES

(Originally published in Modus on November 14, 2019)

Q: How do you know when it's time to leave a job?

You'll have a number of jobs during your career. Some of them will be good. Some of them will be bad. Most of them will oscillate wildly between the two poles. But the one thing all those jobs will have in common is that at some point you will leave. Sometimes it'll be your call, sometimes it won't. Sometimes it'll be a happy occasion, sometimes it won't. But rest assured, every job has an endpoint.

The fact that a job ends shouldn't be a surprise, although the manner in which it ends might be. So let's talk about some of the reasons why people leave jobs and see what we can do to mitigate the negative impact. Because jobs equal income and all that.

But first, let me start with a story. I've been running my own shop (along with my partner Erika Hall, who has a new book about design research out) for 18 years. In those 18 years we've hired somewhere in the neighborhood of 50 people. The first time one of my employees left, I was devastated. I thought it was my fault. I started thinking our little studio sucked, because why else would someone want to leave? I took it personally. It wasn't a good look. I asked them why they were leaving.

"It's just time for something new."

As a former employee myself, that made total sense to me. But in my still-very-newish role as an employer, I'd thrown sense out the window. Here I thought I'd built this great little studio where everyone enjoyed coming to work every morning, which had always been my dream, and someone was leaving. I called up a few colleagues who ran their own small studios. They all told me the same thing:

"Everybody leaves. Get used to it."

As time moved on, and as more people left and others were hired on, I did indeed get used to it. I also came up with a little thing I'd tell new employees on their first day. It went something like this:

> "I'm glad you're here. I hope your stay here is long and fruitful, and that we both get a lot out of it. However, someday you'll want to leave. My most important job is to prepare you for that day, so you can walk out of here well prepared for the next job. While you're here, I'll do my best to teach you everything I know, and the minute you're ready to go, let me know. You'll have my full support on figuring out where you should go next, if you want it. I'll make calls. I'll write references. Leaving is a totally normal part of a job, so let's deal with it like we deal with everything else. We'll plan, we'll research, we'll celebrate. And the sooner you tell me you're thinking of leaving the sooner we can start all that. It'll also give me adequate time to find your replacement."

A few people took me up on that offer, and I kept my word. We researched companies that might be good for them together, I made some calls on their behalf, and I did what I could to get their name out. I even helped a few employees negotiate a better offer from their new workplaces. At the same time, I was interviewing their replacements.

A few employees didn't take me up on the offer, and I understand why they didn't. I imagine every employee has a story about a shitty boss who freaked out when they told them they were leaving. Because I was that boss once. But this is how I believe leaving should be handled: out in the open, with honestly, and as a normal part of work. Sadly, not all bosses understand this, so I get why employees don't want to trust them. If you're in this situation, let me give you a tip: You are not the first person to come up with masking interviews as dental appointments. When you start lining up three dental appointments a week your boss knows what you're doing. Be more clever.

Now that we've established that leaving is an expected and normal part of the job cycle, let's take a look at the reasons people leave.

YOU'RE JUST DONE

Chapters have beginnings, middles, and ends. You can be in a job you totally love, solving problems you're good at solving and enjoy solving, working with people you enjoy working with, and just come in one day and decide you're done. It happens. This job has run its course and it's time to move on to the next thing. If this happens to you, the best thing to do is to take stock of where you are both professionally and personally.

If you're doing project-based work, where in the project cycle are you? If you're close to the end, you should see if you can stick it out. Finish strong. Wrapping up a project is a natural point for closure and you won't leave your coworkers in a bad spot. If you're between projects, it's a great time to leave. If you're in the middle of a project, things get a little trickier. Can you make it to the end? Can you stay focused enough to finish the project with the level of quality people expect from you? If not, be honest with yourself and with your team as soon as possible.

A BETTER OPPORTUNITY CAME UP

Even if you like your current job, you might've been offered something that excites you more. Put some due diligence in. Make sure this isn't a grass-is-always greener situation. Maybe this is the job you've always wanted at the place you've always dreamed of working. You only make this trip through space once. Take it. Just finish strong with the current boss. Nothing ever freaked me out like offering someone a job, asking when they could start, and them replying "I can start right now!" when I knew they already had a job. If someone's willing to ditch their current job without giving notice, they're willing to do it to you.

YOU'VE GROWN BEYOND THE JOB

Sometimes you grow beyond the job you were hired to do. Say, for example, that you were hired to knock out prototypes in Sketch, but you've been learning a ton about design research and you're looking to try your hand at that. If your company doesn't offer opportunities in the new skills you want to practice, you need to convince them to expand. If they're not open to that, you need to find a new place.

YOU'RE IN THE WRONG PLACE

There are big company people and small company people. There are self-motivated people, and there are people who need oversight. Some people like to build things from scratch, some people like to work on mature products. Some people thrive in chaotic environments, some people need a ton of structure. None of these are right, and none of these are wrong. And sometimes you don't know which type of person you are until you find yourself in the wrong environment. The good news is that workplaces of all varieties, sizes, and structures exist. And if you're in the wrong place, you might be making the people around you as miserable as you're making yourself.

YOUR WORKPLACE TURNED OUT TO BE THE HELLMOUTH

I think it's safe to say we're all hopeful when we start a job. We're happy for the opportunity. We're happy to contribute our skills and our labor to society. And obviously, we're happy to be able to earn a living. But sometimes what you thought was going to be a decent place to work turns out to be less than ideal. And look: All workplaces have problems. I'm not talking about the kind of place where someone would reheat fish in the microwave, or ask you to empty your own trash. I'm talking about the kind of place that exploits your labor. The kind of place that would cut your pay right before it had to report quarterly earnings. The kind of place that would use your labor to rise to success, only to abandon you as your efforts were about to come to fruition. The kind of place where management benefits from your labor, but you don't. The kind of place where workers stand up for themselves in order to improve their situation and get punished further. That's a horrible place.

Every worker deserves a safe workplace. Free from harassment. Free from abuse. Free from exploitation. Labor is an exchange. And when it's respected, everyone benefits. The company does well, the workers do well, the people who use the products or services made by that company do well. And finally, the ecosystem in which all that exists does well. But a company that exploits either its workers, or its customers, or the ecosystem in which it operates, doesn't deserve to exist. And it certainly doesn't deserve workers' labor or customers' hard-earned money. Eventually, everybody will leave.

Companies don't have an inherent right to thrive. People do.

THANK YOU

Thanks for reading these essays. Thanks for the support over the years. Thanks for sticking around during this very, very shittiest of times.

Thanks to all the editors, and the proofreaders, an anyone who had to plow through an early draft, as well as anyone who had to plow through a finished draft, for that matter. Thanks to Kat Vellos for her self-publishing advice.

Thanks to the nurses. Thanks to the doctors. Thanks to all the front line workers. Thanks to the postal workers.

Thanks to the community workers. Thanks to the organizers. Thanks to the protestors, the statue demolishers, the cop car flippers, the calm ones, the angry ones, the sad ones, the righteous ones.

Thanks to the people working in grocery stores, who risk their lives every time some jackass decides to pull his mask down to use ApplePay. Thanks to all the delivery drivers. The retail workers. The warehouse workers.

Thanks to the kids making zines again.

Thanks to Black Lives Matter. Thanks to Stacey Abrams. Thanks to The Squad.

Thanks to the union organizers. Thanks to the workers who are out there organizing other workers, trying to get them to understand how incredibly powerful they can be when they band together. Thanks to everyone who understands that you don't work for the person who signs your checks. Thanks to everyone who's about to go here: www.code-cwa.org/

Thanks to the immigrants, and the immigrant grocery stores that extend credit to their communities until payday. Thanks to everone who marched against the Muslim ban. Thanks to everone who marched in support of DACA. Thanks to all the DACA kids doing their thing out there in the community as nurses, doctors, frontline workers, teachers, journalists, firefighters, EMS drivers, doctors, and well... everywhere.

Thanks to Jeff Bezos, Mark Zuckerberg, and Jack Dorsey for making sure none of the rest of us could ever be called the biggest asshole in our industry.

Thanks to our children for not murdering us in our sleep for destroying the planet, and their future along with it.

Thanks to my therapist. Hell, thanks to all the therapists. But especially mine, so works even harder than you can imagine. Lordy.

Thanks to my wife Erika, and my daughter Chelsea.

Thanks to my dog Rupert, may I one day become the good person he believes I already am.

Thanks to everyone who didn't give up.

ABOUT THE AUTHOR

Mike Monteiro is the co-founder and design director of Mule Design, an interactive design studio whose work has been called "delightfully hostile" by The New Yorker, and cashed checks from George Soros. He's written four books—*Design Is a Job, You're My Favorite Client, Ruined by Design,* and the one you're holding now, which you hopefully paid for. He's not as smart as his wife, or as charming as his dog, or as brave as his daughter. He's pretty much given up on tech and is considering a future as a lighthouse keeper or record store owner, or maybe opening a record store in a lighthouse.

Protect trans kids.

ABOUT MULE BOOKS

We are a subsidiary of Mule Design. We write good books to make your life easier. We publish them outside the clownshow of traditional publishing, which exploits authors and disrespects readers.

THANK YOU

Thanks to Rosemary Campbell, Ross Floate, Rod Bebbie, Brian Carr, and everyone who participated in the joke writing but wanted to stay anonymous. I love you, even though you're filthy cowards.

like censoring critics, the case against evolution, a neurosyphilitic South African apartheid stan, an asshole who can only punch down, someone who's so bad at running Twitter that I almost forgot how incredibly shitty Jack Dorsey was at running Twitter, a guy who took HGH and it didn't make him look strong but did make his ribcage into a weird shape, a human paraquat, Reddit personified, someone so wretched that SnuffleupagUs is still hiding from him, the sort of landowner who gives you fond feelings about Mao, a guy whose neck was made for the sweet embrace of a cold guillotine blade, a free-speech warrior who sues companies that take advantage of the right to not advertise on his shitty social media site, the sentient comment section of a red pill forum, the only acceptable argument for the phrase "go back to Africa", a hateful motherfucker that even the world's sweetest dog would snap at, a man so vile that if he and Joe Rogan were hanging off the edge of a cliff and I were forced to save one of them—and trust that I would absolutely hate myself for this—I would save Joe Rogan, a venereal disease dispensery, someone whose touch I imagine is so revolting and oddly moist that if I mistakenly brushed up against him I'd hack my own skin off, someone who definitely took a shit on Epstein's jet, someone so white that his asshole makes anal bleach uncomfortable, someone so white the Kansas City Chiefs offered him a kicking contract, someone so white even Lana Del Rey was too uncomfortable to write a song about him, someone so vile that last time he went to church Jesus climbed down off the cross and whipped his ass, someone so disgusting that even Jimmy Fallon thought twice about tousling his hair I mean he eventually went ahead and did it because he's Jimmy Fallon but he thought twice about it, human phlegm, a milky white wet bus seat, an aberration of such vile Biblical proportions that we made sure this sentence was 1,289 words long—exactly one word longer than Faulkner's record for longest run-on sentence so that if you google "longest sentence ever published," this is the motherfucking sentence that comes up, and a real fucking loser, walk into a bar.

The bartender wanders over and says "Hey, aren't you Elon Musk?"

like a real affront to the proud upstanding cuck community—but Chelsea Manning did definitely fuck his girlfriend though, A Geordie Greep song protagonist, who genuinely feels there needs to be more white babies in the world, a man who can make Peter Thiel look like the comparatively well-adjusted person on the list of PayPal founders, a toddler dressed in his father's suit, a car-company owner that would make Henry Ford say "Ooh! That opinion's a little spicy! Maybe keep it to yourself!", a man who tries to impress nerds by lying about his achievements playing video games, a real cock—turgid, erumpent, and vascular—but not in a fun way, a man who somehow gives sons of White South African emerald mine owners a bad name, an utter waste of skin, a shitty eleven-year-old boy who wished he was Big to a Zoltar machine on a pier, a toilet seat that smokes a cigar, a sniveling dickwad poopy-pants fuckface, a man who makes Trump look like the normal one in a photo of them on stage at a rally, a man who doesn't have the intellectual ability to see any historical reason why encouraging the rise of far-right parties in German and Italy might lead to global problems, a sort of secular year-round beige Grinch, a man who got kicked out of OpenAI for being a bit too crazy for Sam Altman to deal with, a malfunctioning automaton from Robber Baron Westworld, a man who had no part in the founding of Tesla but throws a big pissy-cry-baby fit if he isn't described in the press as a founder, a man who spends most of his energy trying to impress basement-dwelling troglodyte man-children online, the weird foreign exchange kid standing in the corner of homeroom staring at you creepily who you're desperately hoping won't ask you to prom, a wet fart of a man, the worst tech billionaire which is really saying something, a billionaire who makes you nostalgic for the good-old-days of 2008 when the worst money-and-status-obsessed drug-addicted shitstain humans destroying the country ran banks, someone that Santa Claus would curb-stomp, an enormous fucking dork, the kind of guy who would take sildenafil and it would just make him taller, the posterboy for money not bringing you happiness, a permanent fixture in everyone's nightmare blunt rotations, a man whose jokes are so deeply unfunny that he had to buy an entire social media platform to get anything more than a "haha, cool" from his most loyal sycophants, a gronk whose meme-gagging drug-addled validation-seeking love of attention from Nazi-adjacent weebs has metastasized so far that Shiba Inu dogs are embarrassed to show their curly tails on the street, a Wired magazine centerfold, a guy who seems genuinely, like a child, or a medium size lizard, but with a smoother brain, into straight up fascist dictator shit

A white supremacist, a TERF, a bad father, the world's biggest loser, a guy who really puts the "hole" in K-hole, a mining tycoon, an eproctophile, a serial toxic podcast guest, a eugenicist, a vivisectionist, a sociopathic narcissist, a union buster, a horrible boss, a nepo baby, a deeply insecure man, a narcissist born on third base who celebrates the triple he didn't hit and insists he was robbed of a home run, an adult who still laughs at "420" and "69", someone who thinks he's funny but he's not, an edgelord, a vile pustule of rotting flesh, an intolerant tantrum-throwing dickweasel, someone who's been handed everything and achieved nothing on his own, someone who didn't get enough love as a child, someone who stayed up all night (on ketamine) writing a sink joke, an undocumented immigrant illegally overstaying his visa, a man incensed by the idea that his father owned the emerald mine that he absolutely owned, a man calling random other people a "pedo guy," someone whose own father couldn't stand him, a guy who tried to buy off an employee he exposed himself to with a pony, a guy who is convinced his hair transplant is convincing, a guy who named the models of his shitty cars so they'd spell S-E-X, a failson, a bloke who called the Thai cave rescue hero a paedophile, someone who said he could cure world hunger but didn't want to, a guy who cries against gender affirming care but totally got hair plugs, a guy who Peter Thiel has been dunking on for over 20 years, a guy who definitely gave a Nazi salute multiple times, someone who still thinks "that's what she said" is funny, a forgotten and disappointing Bond villain that even George Lazenby could take out in twelve minutes flat, a shitty video game player who pays others to grind out in-game treasure and status so he can pretend to be legit, a dude who's spoken at Nazi rallies, a guy who called himself "dark" and "gothic" with a weird smile at a political rally, a toxic male, a previously only theorized twenty-fifth season of Family Guy somehow given sapience, a man with all the charm of something Divine would eat off the ground, someone who made his midlife crisis everyone's problem, someone who ordered "all boys" at the fertility clinic, the owner of the exploding shitty truck factory, a grown man who thinks the letter "X" is inherently cool, a man with a breeding kink, the sort of fellow I'd call a cuck if that didn't feel

The Official
Elon Musk
Joke Book

A NEW STANDARD OF HUMOR

OBLIVION BROS.

The Official Elon Musk Joke Book

A VERY SATISFYING GROUP PROJECT

MIKE MONTEIRO

www.ingramcontent.com/pod-product-compliance
Lightning Source LLC
Chambersburg PA
CBHW010330030426
42337CB00026B/4881